SpaceX

Making Commercial Spaceflight a Reality

Erik Seedhouse

SpaceX

Making Commercial Spaceflight a Reality

 Springer

Published in association with
Praxis Publishing
Chichester, UK

Dr Erik Seedhouse, M.Med.Sc., Ph.D., FBIS
Milton
Ontario
Canada

SPRINGER–PRAXIS BOOKS IN SPACE EXPLORATION

ISBN 978-1-4614-5513-4   E-ISBN 978-1-4614-5514-1
DOI 10.1007/978-1-4614-5514-1

Springer Dordrecht Heidelberg London New York

Library of Congress Control Number: 2013941041

Springer Science+Business Media New York © 2013

Cover design: Jim Wilkie
Project copy editor: Christine Cressy
Typesetting: BookEns, Royston, Herts., UK

Springer is a part of Springer Science + Business Media (springer.com)

Contents

Preface

January 23rd, 2012, marked the start of the Year of the Dragon in the Chinese calendar and, in May 2012, SpaceX's Dragon became the first privately developed spacecraft to visit the International Space Station (ISS). Space travel is one of the most difficult of all human endeavors, and success is never guaranteed. The Dragon flight introduced a series of new challenges and new magnitudes of complexity and, by docking with the ISS, SpaceX once again made history by becoming the first private company to send a spacecraft to the ISS.

Dragon is a spacecraft unlike any other. Not only is it the first privately developed spacecraft to successfully return from Earth orbit, it is also the only reusable spacecraft in operation today. It also happens to be just another element in Elon Musk's goal of making humanity a spacefaring civilization. Just as Elon Musk's PayPal product took internet payments that cost US\$0.40 or more per transaction via credit cards and made them free, his SpaceX rockets and spacecraft are going to revolutionize space travel with new lower prices. While humanity becoming a spacefaring species may be inevitable in the long term, if personal income keeps growing, applying modern manufacturing, testing, control, and management techniques to spaceflight may allow us to see substantial strides this decade. Leading the charge will be SpaceX.

SpaceX is applying modern manufacturing techniques such as friction stir welding and modern CAD and production data management techniques to build its rockets. It's also developing its Falcon 1, 9, and other rockets in quick succession, reusing many components and design and manufacturing strategies. Not satisfied with business as usual, SpaceX doesn't rely on decades-old space-proven products or even the veteran aerospace testing firms; instead, it builds new components and tests them in-house.

SpaceX: Making Commercial Spaceflight a Reality is an account of commercial spaceflight's most successful start-up. It describes the extraordinary feats of engineering and human achievement that have placed SpaceX at the forefront of the launch industry and positioned it as the most likely candidate for transporting humans to Mars. Since its inception in 2002, SpaceX has sought to change the space launch paradigm by developing a family of launch vehicles that may ultimately reduce the cost and increase the reliability of space access by a factor of 10.

This book describes how SpaceX is based on the philosophy that simplicity, low cost, and reliability *can* go hand in hand. It explains how, by eliminating the traditional layers of management, internally, and subcontractors, externally, SpaceX reduces its costs while speeding decision-making and delivery. Likewise, by keeping the vast majority of manufacturing in-house, the book explains how SpaceX reduces its costs, keeps tighter control of quality, and ensures a tight feedback loop between the design and manufacturing teams.

Forged by Elon Musk in 2002, the founder of PayPal and the Zip2 Corporation, SpaceX has already developed two of the coolest new launch vehicles, established an impressive launch manifest, and been awarded funding by NASA to demonstrate delivery and return of cargo to the ISS. Supported by this order book and Mr. Musk's substantial resources, SpaceX is on an extremely sound financial footing as the company moves towards volume commercial launches.

Although drawing upon a rich history of prior launch vehicle and engine programs, SpaceX is privately developing the Dragon crew and cargo capsule and the Falcon family of rockets from the ground up, including main and upper-stage engines, the cryogenic tank structure, avionics, guidance and control software, and ground support equipment. The Falcon 9 and Falcon Heavy are the only US launch vehicles with true engine out reliability. They are also designed such that all stages are reusable, making them the world's first fully reusable launch vehicles. And the Dragon crew and cargo capsule, currently under development, may revolutionize access to space by providing efficient and reliable transport of crew and cargo to the ISS and other low Earth orbit destinations. This book explains how. Here is an up-close portrait of the maverick company that is, in short, one of the most spectacular aviation triumphs of the twenty-first century.

Acknowledgments

In writing this book, the author has been fortunate to have had five reviewers who made such positive comments concerning the content of this publication. He is also grateful to Maury Solomon at Springer and to Clive Horwood and his team at Praxis for guiding this book through the publication process. The author also gratefully acknowledges all those who gave permission to use many of the images in this book, especially Hannah Post at SpaceX Media Relations.

The author also expresses his deep appreciation to Christine Cressy, whose attention to detail and patience greatly facilitated the publication of this book, and to Jim Wilkie for creating the cover.

To
Julian

About the author

Erik Seedhouse is a Norwegian-Canadian suborbital astronaut whose life-long ambition is to work in space. After completing his first degree in Sports Science at Northumbria University, the author joined the legendary 2nd Battalion the Parachute Regiment, the world's most elite airborne regiment. During his time in the "Para's", Erik spent six months in Belize, where he was trained in the art of jungle warfare. Later, he spent several months learning the intricacies of desert warfare on the Akamas Range in Cyprus. He made more than 30 jumps from a Hercules C130 aircraft, performed more than 200 abseils from a helicopter, and fired more light anti-tank weapons than he cares to remember!

Upon returning to the comparatively mundane world of academia, the author embarked upon a Master's degree in Medical Science at Sheffield University. He supported his studies by winning prize money in 100-km running races. After placing third in the World 100 km Championships in 1992 and setting the North American 100-km record, the author turned to ultradistance triathlon, winning the World Endurance Triathlon Championships in 1995 and 1996. For good measure, he also won the inaugural World Double Ironman Championships in 1995 and the infamous Decatriathlon – an event requiring competitors to swim 38 km, cycle 1,800 km, and run 422 km. Non-stop!

Returning to academia in 1996, Erik pursued his Ph.D. at the German Space Agency's Institute for Space Medicine. While conducting his Ph.D. studies, he still found time to win Ultraman Hawai'i and the European Ultraman Championships as well as completing the Race Across America bike race. Due to his success as the world's leading ultradistance triathlete, Erik was featured in dozens of magazines and television interviews. In 1997, *GQ* magazine nominated him as the "Fittest Man in the World".

In 1999, Erik decided it was time to get a real job. He retired from being a professional triathlete and started his post-doctoral studies at Vancouver's Simon Fraser University's School of Kinesiology. In 2005, the author worked as an astronaut training consultant for Bigelow Aerospace and wrote *Tourists in Space*, a training manual for spaceflight participants. He is a Fellow of the British Interplanetary Society and a member of the Space Medical Association. In 2009, he was one of the final 30 candidates in the Canadian Space Agency's Astronaut Recruitment Campaign. Erik works as a manned spaceflight consultant, professional speaker, triathlon coach, and author. His spaceflight company, Suborbital Training (*www.suborbitaltraining.com*), provides customized training programs for commercial suborbital astronauts and tourists. He is the Training Director for Astronauts for Hire (*www.astronauts4hire.org*) and completed his suborbital astronaut training in May 2011. Between 2008 and 2012, he served as director of Canada's manned centrifuge operations.

In addition to being a suborbital astronaut, triathlete, centrifuge operator, and director, pilot, and author, Erik is an avid mountaineer and is currently pursuing his goal of climbing the Seven Summits. *SpaceX* is his eleventh book. When not writing, he spends as much time as possible in Kona on the Big Island of Hawaii and at his real home in Sandefjord, Norway. Erik and his wife, Doina, are owned by three rambunctious cats – Jasper, Mini-Mach, and Lava.

Abbreviations and acronyms

ARIS	Active Rack Isolation
ARRA	American Recovery and Reinvestment Act
ARS	Air Revitalization System
ASIL	Avionics Software Integration Laboratory
ATK	Alliant Techsystems
ATV	Automated Transfer Vehicle
C3PO	Commercial Crew and Cargo Program Office
CAM	Collision Avoidance Maneuver
CBM	Common Berthing Mechanism
CCDev	Commercial Crew Development
CCiCap	Commercial Crew Integrated Capability
CCP	Commercial Crew Program
CDR	Critical Design Review
CIR	Combustion Integrated Rack
COTS	Commercial Orbital Transportation Services
CRS	Commercial Resupply Services
CST	Commercial Space Transportation
DARPA	Defense Advanced Research Projects Agency
ECLSS	Environmental Controlled Life Support System
EDS	Emergency Detection System
EELV	Evolved Expendable Launch Vehicle
ESA	European Space Agency
FAA	Federal Aviation Administration
FDM	Free Drift Mode
FRR	Flight Readiness Review
GNC	Guidance Navigation and Control
GPS	Global Positioning System
GTO	Geosynchronous Transfer Orbit
ICD	Interface Control Document
IMU	Inertial Measurement Unit
ISBR	Integrated System Baseline Review

ISS	International Space Station
JSC	Johnson Space Center
KSC	Kennedy Space Center
LAS	Launch Abort System
LCPE	Low Cost Pintle Engine
LEM	Lunar Excursion Module
LEO	Low Earth Orbit
LIDAR	Light Detection and Ranging
LLM	Liberty Logistics Module
LMLE	Lunar Module Landing Engine
LOX	Liquid Oxygen
LRR	Launch Readiness Review
LVA	Launch Vehicle Adapter
MDA	McDonald Dettweiler and Associates
MPCV	Multi-Purpose Crew Vehicle
MSRR	Materials Science Research Rack
NERVA	Nuclear Engine for Rocket Vehicle Applications
OMAC	Orbital Maneuvering and Attitude Control
OSC	Orbital Sciences Corporation
PAF	Payload Attach Fitting
PCM	Pressurized Cargo Module
PDR	Preliminary Design Review
PICA	Phenolic Impregnated Carbon Ablator
RGPS	Relative Global Positioning System
SAA	Space Act Agreement
SDS	Spacecraft Docking System
SHERE	Shear History Extensional Rheology Experiment
SLS	Space Launch System
SNC	Sierra Nevada Corporation
SRB	Solid Rocket Booster
SRB	Safety Review Board
SRR	System Readiness Review
SSC	Stennis Space Center
SSME	Space Shuttle Main Engine
TEA	Triethylaluminum
TEB	Triethylborane
TIM	Technical Interface Meeting
TRL	Technology Readiness Level
TVC	Thrust Vector Control
UHF	Ultra High Frequency
ULA	United Launch Alliance
USAF	United States Air Force
VAIL	Vehicle Avionics Integration Laboratory
VTHL	Vertical Take-off Horizontal Landing
VTVL	Vertical Take-off Vertical Landing

1

Elon Musk: The space industry's Tony Stark

After a near flawless nine-day mission, the Dragon capsule splashed down on target in the Pacific Ocean just off the coast of Mexico, marking the end of the first commercial mission to ferry supplies to the International Space Station (ISS). Tethered to three large parachutes, the unmanned gumdrop-shaped capsule (Figure 1.1), which had carried food, water, clothing, and equipment to the orbiting outpost, hit the water at 8:42 a.m. local time on May 31st, 2012, about 900 kilometers west of Baja, California, witnessed by technicians from the remarkable company that had built and flown it – Space Exploration Technologies, or SpaceX:

1.1 Dragon capsule. Courtesy: SpaceX

1.2 Elon Musk. Courtesy: NASA

"This really couldn't have gone better. I'm overwhelmed with joy. It's been 10 years, and to have it go so well is incredibly satisfying."

Elon Musk, SpaceX chief executive, speaking at a televised
news conference from the company's headquarters
in Hawthorne, California

The Dragon had separated from the ISS about seven hours before splashdown, after astronauts had loaded it with used equipment, experiment samples, and trash. The success of what was really just a trial run for the spacecraft positioned SpaceX to begin regular supply missions with bigger payloads to the ISS and paved the way for manned missions perhaps as early as 2015. The flight of the Dragon was also notable for the fact that, since the Space Shuttle program had ended the previous year, the station had been resupplied by Russian and European spacecraft.

In 2002, Elon Musk (Figure 1.2) was just another Internet mogul starting a commercial space company. But Musk was bolder than his peers. Simply providing a suborbital trip to space like Sir Richard Branson's *SpaceShipOne*[1] wouldn't satisfy the South African native; Musk wanted to fly resupply missions with astronauts to the ISS.

---

[1]   That rocket, and the passenger version that will make up Sir Richard Branson's Virgin Galactic fleet reached an altitude of more than 100 kilometers in 2004.

It was a bold goal because, as any space engineer will tell you, getting to orbit is by several orders of magnitude more difficult than reaching suborbital altitudes. In fact, it is such a challenge that only eight countries and a few private companies have reached orbit independently. Orbital flight also happens to be very, *very* expensive, but Musk reckoned he could do it cheaper *and* turn a profit. His plan? Run his company like an Internet start-up and launch a new age in space exploration.

Perhaps one of the most intriguing aspects about how Musk works is the fact that he works at all. After all, by his early thirties, his Internet ventures had made his net worth about US$200 million. He could have retired but chose instead to enter perhaps the riskiest, costliest, and most unforgiving businesses there is: launching rockets.

Born in South Africa in 1971, the son of a Canadian mother and a South African father, it didn't take long for Musk to demonstrate his entrepreneurial spirit. He bought his first computer at the age of 10 and quickly taught himself computer programming. Two years later, he wrote code for Blastar, a video game, which he subsequently sold to a computer magazine for US$500. Then, when he was 17, spurred by the prospect of avoiding compulsory service in the South African military,[2] Musk moved to Canada, spending two years at Queen's University, Kingston. He had planned a career in business and worked at a Canadian bank one summer as a college intern. After Kingston, Musk moved to the US, where he earned degrees in physics and business at the University of Pennsylvania. He had intended to begin a graduate program at Stanford in 1995 but, after just two days, chose instead to devote the next four years to developing Zip2, a company that enabled companies to post content on the Internet. In February 1999, Compaq Computer Corporation bought Zip2 for US$307 million – in cash. It was one of the largest cash deals in the Internet business at the time and Musk walked away with a cool US$22 million for his 7% share. He was only 28.

He used US$10 million to start X.com, an online bank, which went online in December 1999. The following month, Musk married his first wife, Justine, whom he had met while studying in Canada. Two months later, in March 2000, X.com merged with Confinity, which had developed a service you may have heard of – PayPal, which provides customers with payment transactions over the Internet. Musk increased his fortune when eBay bought PayPal for US$1.5 billion in 2002. The deal saw his net worth rocket past US$100 million. By that time, he and Justine had moved to Los Angeles and had their first child, a boy named Nevada Alexander. Tragically, while having a nap one day, Musk's son stopped breathing and, by the time the paramedics had resuscitated him, the 10-week-old infant had been without oxygen for so long that he was pronounced brain-dead. He spent three days on life

---

[2]   Musk has explained in several interviews that he doesn't have a problem with military service but he didn't like what the South African military was doing in the late 1980s, especially the brutal oppression of the black majority. When he moved to Canada to avoid conscription, it was against the wishes of his father and the two rarely speak as a result of the younger Musk's decision.

support before Musk and his wife made the agonizing decision to take him off it. Sudden Infant Death Syndrome was the verdict.

Having had enough of the Internet, Musk searched for a new challenge and founded Space Exploration Technologies, or SpaceX, in June 2002. To kick-start his company, he tried buying a rocket from Russia, but soon realized the proposition was too risky and instead considered building his own rocket. Establishing a rocket company was seen by many in the space industry as an audacious move. After all, Musk possessed very little background in the field of rocket science. He could have been forgiven if he had chosen to buy rockets from established rocket-building companies but that just wouldn't have been Musk. Instead, he decided to build SpaceX from the ground up. His initial goal was to reduce the cost of launch services – a milestone spurred by Musk's frustration with not only how much money NASA spent on the space program, but also how little the costs of space exploration have decreased in the decades since the end of the Apollo Program. Once he had solved the inefficiencies of the space program, Musk had his sights set on low-cost human travel into orbit and establishing a colony on Mars. But, before he could send humans to Mars, Musk needed to get his rockets into orbit.

The challenges facing Musk were formidable. Between 1957 and 1966, just as the space age was gaining momentum, the US had sent 429 rockets into orbit, a quarter of which failed. To date, only governments have managed to harness the capital and intellectual muscle necessary to launch rockets into orbit. In fact, practically every Russian, Chinese, and American rocket that exists today is a legacy of ballistic missiles. And building those rockets didn't come cheap. The American, Russian, and Chinese space programs required small armies of engineers working with nearly unlimited budgets. For example, the Apollo Program employed more than 300,000 people and cost more than US$150 billion in 2007 dollars, or more than 3% of the US federal budget. Even the now-retired Space Shuttle required a ground crew of 50,000 and cost more than half a billion dollars *every* time it flew. Incidentally, even the extraordinary amounts of money that were thrown at the Shuttle didn't increase safety because it is still the most dangerous rocket system ever created. (NASA administrators originally stated the risk of catastrophic failure was around one in 100,000; NASA engineers put the number closer to one in 100; a more recent report from NASA said the risk on early flights was one in nine. The eventual failure rate was two out of 135.)

The few private companies that *have* managed to get something into orbit have used hardware developed under government programs. And their services aren't cheap. If you want to launch a satellite into orbit on a Sea Launch Zenit (Figure 1.3), it will set you back a cool US$50 million to US$75 million. Even if you happen to be the US Air Force (USAF), putting a 200-kilogram payload into low Earth orbit (LEO) on an Orbital Sciences *Pegasus* will cost around US$30 million.

"What's the fastest way to become a commercial space millionaire? Start as a commercial space billionaire."
                    Hackneyed joke spawned by the number of companies that
                    have tried and failed to launch rockets into LEO

1.3 Sea Launch Zenit. Courtesy: Sea Launch

To Musk, launch prices were a damning indictment of the state of space exploration, a business that had spent hundreds of billions of dollars on rocket technology in the past 50 years with the result that, before SpaceX came along, the cost of putting something into LEO was still around US$10,000 per pound. It was this lack of progress that particularly frustrated Musk, who decided he would aim to reduce those costs by half – or more.

To many space industry observers, it was a tall claim.

Musk knew the stakes would be high. After all, he knew very little about the rocket industry and had never actually built anything – except Internet companies – in his life. The odds were hardly in his favor. But Musk had thrived in businesses where the default expectation was failure, so why not roll the dice on building rockets? The question was how to do it.

Musk started by going to the heart of the aerospace world in El Segundo, California, one of the beach cities just south of Los Angeles International Airport, and began recruiting industry veterans for SpaceX. One of his first hires was Tom Mueller, one of the world's leading propulsion experts. Designing propulsion systems had come naturally to Mueller, who came from a hands-on background. Born in St Maries, Idaho, a tiny logging community of about 2,500 people, Mueller's dad was a log truck driver and he wanted his son to be a logger, so it was only natural that the younger Mueller grew up around logging trucks and chainsaws. It was an environment that spurred an interest in figuring out how things work, which probably explains why he took his dad's lawnmower apart. When his dad found the parts, he was upset because he didn't think he could put the pieces back together, but the younger Mueller reassembled it and the machine ran pitch perfect. From rebuilding lawn mowers, Mueller moved on to building and flying toy rockets. He bought Estes rockets from his local hobby shop, although they didn't last long because he usually crashed them or blew them up. In junior high, Mueller submitted a hybridized life sciences propulsion project to the science fair which was to fly an Estes rocket with crickets in it to see what the effects of acceleration were on the crickets. Unfortunately, the parachute failed and the deceleration when the rocket hit the ground killed the crickets. Not wanting to kill any more wildlife, Mueller restricted his next project to building a rocket engine out of his dad's oxyacetylene welder and made a rocket engine out of it by injecting water into it to see what effect that had on its performance. The first time he ran it, the engine burned a hole through the side of the chamber but, with some minor modifications, he was able to run it in a steady state – an achievement that allowed him to reach the regional round of the science fair.

Mueller earned a master's degree in mechanical engineering from the Frank R. Seaver College of Science and Engineering and received job offers for work in Idaho and Oregon, but it was for non-rocket stuff. So Mueller decided to move to California to get a rocket job, eventually taking a position with TRW Space and Electronics, where he spent 14 years running the Propulsion and Combustion Products Department. Along the way, he earned the TRW Chairman's Award and filed several patents in propulsion technology. Mueller was happy working there, but his ideas about rocket engine design in a company in which rocket engines weren't a core component were lost. To satisfy his creative impulses, Mueller turned to the Reaction Research Society, building his own engines and launching them in the Mojave Desert with fellow rocketeers.

Even for an experienced propulsion engineer like Mueller, building rocket engines wasn't easy; these engines rely on myriad valves and seals to control the flow and need super-cooled oxidizers to mix with the fuel so it can ignite. The resulting combustion, which can be described as a controlled explosion, is channeled at high pressure into the nozzle, creating the thrust that propels the rocket. These challenges didn't deter Mueller though. By 2002, he had almost completed the world's largest amateur liquid-fuel rocket engine, capable of producing 13,000 pounds of thrust. Musk met the enterprising propulsion engineer in January 2002 just as Mueller was preparing to attach his monster engine to an airframe. For Musk, building rocket engines was the key to his commercial spaceflight enterprise. He took one look at the rocket engine and asked whether Mueller could build a bigger one.

1.4 Gwynne Shotwell was recruited by CEO Elon Musk in 2002 to be vice president of business development. Shotwell is responsible for day-to-day operations of SpaceX and manages customer and strategic relations, including those with NASA. Courtesy: NASA

Having recruited Mueller and a slew of other rocket engineers, Musk needed someone to run day-to-day operations.

Gwynne Shotwell (Figure 1.4) ran into Musk when she dropped off a friend from lunch who had just started working at SpaceX. The friend had mentioned to Musk that he should hire a business developer and Musk agreed, hiring Shotwell, SpaceX's seventh employee, as Vice President of Business Development. In that position, she built the Falcon vehicle family manifest to over 40 launches, representing over US$3 billion in revenue. Today, as president, Shotwell is the powerhouse of the company, responsible for day-to-day operations and for managing all customer and strategic relations to support company growth.

Prior to joining SpaceX, Shotwell spent more than 10 years at the Aerospace Corporation, where she held positions in Space Systems Engineering & Technology and Project Management. She was promoted to the role of Chief Engineer of a medium launch vehicle (MLV)-class satellite program, managed a landmark study for the Federal Aviation Administration (FAA) on commercial space transportation, and completed an extensive analysis of space policy for NASA's future investment in space transportation. Shotwell was subsequently recruited to be Director of Microcosm's Space Systems Division, where she served on the executive committee and directed corporate business development.

With his team complete, all Musk had to do was to get on with the business of building rockets and rocket engines. Traditionally, rocket manufacturers bought engines from established companies because the prospect of designing and building your own rocket engines ... well, that would be madness. But that is exactly what Musk intended to do: he would make his propulsion systems in-house.

Not content with the challenges of establishing a rocket company, Musk also took time out to make his cae for commercial spaceflight to the Senate (Appendix I) and co-founded Tesla Motors.[3] One of the main objectives of Tesla Motors (named after electrical engineer and physicist Nikola Tesla) was to develop an environmentally friendly sports car. To that end, the company built the snappy Tesla Roadster (Figure 1.5) – a car that charges overnight, uses no gasoline, and sprints from zero to 96 kilometers per hour in less than four seconds.

Tesla Motors, a public company that trades on the NASDAQ stock exchange under the symbol TSLA, gained widespread attention by producing the Roadster, but the company, while still successful, wasn't the dramatic success of Musk's previous business ventures. Based in Palo Alto, California, and employing almost 900 full-time employees, Tesla sells its patented electric power-train components to

---

[3]   The genesis of the Tesla enterprise occurred in 2003 when two teams, one consisting of Martin Eberhard and Marc Tarpenning and the other of Ian Wright, J.B. Straubel, and Musk were trying to think of ways to commercialize the T-Zero prototype electric sports car created by AC Propulsion. One of Musk's goals had always been to commercialize electric vehicles, starting with a premium sports car before moving to mainstream vehicles. It was suggested the teams join forces to maximize the chances of success, so Musk became chairman, Eberhard took on the role of CEO, and Straubel became CTO.

1.5 The Tesla Roadster is an all-electric sports car, 2,250 of which have been sold in 31 countries through March 2012. The car, which costs US$109,000, can accelerate from zero to 60 miles per hour (100 kilometers per hour) in under four seconds and has a top speed of 200 kilometers per hour. Courtesy: Wikipedia

other car manufacturers so they can deliver their own electric vehicles (EVs) to customers sooner. For example, Tesla builds electric power-train components for the Smart urban commuter car, the Mercedes A-Class hatchback, and the lowest-priced Toyota SUV, the RAV4.

While Musk was developing the Tesla venture, his SpaceX team had started building a two-stage rocket powered by "Merlin", a compact, durable engine designed to lift a first stage, and "Kestrel", the second. The rocket, christened Falcon 1 (Figure 1.6), was designed to lift 1,400 pounds into LEO. Getting the payload into orbit was to prove a tougher task than Musk had envisaged.

First there was the issue of finding a launch site. Musk had originally hoped to launch the Falcon 1 booster (carrying a TacSat-1 satellite built by the US Naval Research Laboratory) from the Vandenberg Air Force Base in California, but that was stymied by a delay launching a Titan-4 rocket carrying a classified payload. After spending an estimated US$7 million on its Vandenberg Air Force Base facilities, SpaceX was told to get out of the Complex 3 West launch site. "It is just, I think, a travesty," Musk told SPACE.com in an interview at the 19th Annual Conference on Small Satellites. Musk, having signed an agreement with USAF to use Complex 3 West and having made investments in the site as well as having paid for the requisite environmental assessments, had cause to be annoyed.

Fortunately, SpaceX had an alternate launch site on Omelek Island in the Kwajalein Atoll, a location that had been part of SpaceX's plans to orbit payloads from there as well as from California and Florida. Kwajalein Atoll, which is part of

1.6 Falcon 1 launch. Courtesy: SpaceX

the Republic of the Marshall Islands (RMI), lies in the Ralik Chain, 3,900 kilometers south-west of Honolulu, Hawaii. On February 6th, 1944, the atoll was claimed by the US and was taken, with the rest of the Marshall Islands, as a Trust Territory of the US. In the years following the American invasion, the atoll was converted into a staging area for further campaigns in the advance on the Japanese homeland in the Pacific War and as a command center for Operation Crossroads and for nuclear tests at the Marshalls atolls of Bikini and Enewetak. The atoll is controlled by the US military under a long-term lease and is part of the Ronald Reagan Ballistic Missile Defense Test Site.

Falcon 1 was shipped to Omelek Island by barge for a projected launch date of September 30th, 2005, although Musk acknowledged delays were likely. While Omelek Island was remote, the location offered some advantages. To begin with, there are no population centers nearby, making range safety easier and, secondly, just about any orbit is achievable from Kwajalein, thanks to it being so close to the equator. Thirdly, work at Kwajalein only had to be done to satisfy one entity – the Environmental Protection Agency – whereas, in California, multiple federal agencies had to be engaged, along with state and county entities. The downside to launching from Omelek was the problem of lugging all the equipment needed to launch a rocket into space to such a remote site, and the challenge posed by the humidity, temperature, and sea spray, which, when combined, created just about the most corrosive environment on the planet. It was this last factor that was to prove costly.

At about the same time as Falcon I was being shipped to Omelek Island, Musk, despite not having flown a single rocket, had already signed three launch contracts (with the Swedish Space Corporation, MacDonald, Dettweiler and Associates Ltd (MDA) of Canada, and a commitment by an unspecified US company) and had invested about US$100 million in SpaceX. There was a lot riding on the first launch.

The 21-meter-high Falcon 1 rocket was the first in a family of boosters planned by SpaceX to offer a more affordable option to launch satellites. Cost-capped at just US$6.7 million, Falcon 1 launches were designed to carry up to 570 kilograms into LEO. Fueled by kerosene and liquid oxygen, the booster featured SpaceX's in-house-designed Merlin engine and a reusable first stage, which, if everything went according to plan, would parachute back to the ocean for later recovery for use on future flight.

Falcon 1's first launch attempt came on November 26th, 2005 at 5:11 p.m. Musk played down the chances of success, telling reporters that the likelihood of a new rocket launching from a new launch pad on its first attempt was low. Musk's prediction proved right on the mark when the launch had to be scrubbed when an auxiliary liquid-oxygen (LOX) fill tank had a manual vent valve incorrectly set to vent. The time it took to correct the problem resulted in too much LOX boiling off and too much helium was lost. Since LOX was used to chill the helium bottles, the loss of the LOX meant there was no way to cool the bottles. The launch team managed to refill the LOX tanks, but the rate at which they could add helium was slower than the rate at which LOX was boiling away, so the decision was taken to scrub.

Launch attempt #2 took place on December 19th, 2005, but a structural issue with the first-stage fuel tank (a faulty valve caused a vacuum in the first-stage fuel tank which sucked inward and caused structural damage) resulted in another scrub. After replacing the first stage, the launch of Falcon 1 was set to take place on February 10th, 2006. Despite the first two scrubs, Musk was unfazed, telling reporters that hiccups were to be expected with the debut of any new launch system, adding that each launch attempt brought valuable experience to the flight team. True enough, but it was probably experience that the flight team was growing weary of, especially when launch attempt #3 resulted in a scrub after problems cropped up during a planned engine test.

Saturday, March 25th, 2006, was the date set for launch attempt #4. This time, the launch proceeded according to plan. One minute before launch, Falcon 1 switched to the computerized launch sequence. Seconds before launch, a spark lit, which fired a turbo-pump spinning at 21,000 rpm, pushing LOX and kerosene into the Falcon's main engine combustion chamber. A moment later, flames erupted beneath the Falcon, generating 72,000 pounds of thrust. That was as good as it got. Immediately following lift-off, the vehicle rolled, rocking back and forth, and then, at T+26 seconds, rapidly pitched over. Impact occurred at T+41 seconds onto a dead reef less than 100 meters from the launch site. The FalconSAT–2 payload separated from the booster and landed on the island, sustaining minor damage.

"We had a successful liftoff and Falcon made it well clear of the launch pad, but unfortunately the vehicle was lost later in the first stage burn," Musk said in an update posted to the company's website. Shortly after the launch, Musk, accompanied by Mueller, the range safety officer, and the vice presidents of avionics and structures boarded a helicopter and flew over Omelek. Except for a fuel slick just offshore and a few scattered pieces of debris, there was little left of the rocket. Four years, tens of millions of dollars, and endless seven-day work weeks. For the engineers, many of whom had quit steady jobs with Boeing and Lockheed, it wasn't the pay-off they had been hoping for.

After poring over video footage, data points, and flight telemetry, the cause of the launch failure was identified as a small fire that had broken out on the first-stage engine. The fire had been caused by a fuel leak and the cause of the fuel leak was an aluminum nut from the fuel pump – the nut had cracked, having corroded in the salty, humid air. The choice to go with aluminum fittings rather than more durable stainless steel had been a cost-saving decision: in the business of launching rockets, weight equals money and that meant choosing aluminum over stainless steel. Unfortunately, that aluminum had been sitting in the humid tropical environment for 10 weeks, with predictable and costly results.

Fixing the nut corrosion problem was simple: SpaceX replaced the aluminum with stainless steel. The team also added fireproof baffling around the engines and, as a further precaution against the heavy tropical air, kept the rocket inside a Quonset hut until a few days before lift-off. They also updated the launch software. Computers had recorded the fuel leak that destroyed the first Falcon 1, but nobody had noticed, so the new launch system was devised to automatically abort a countdown in the event of an anomaly. In all, engineers eventually made 112 changes to the rocket and the launch sequence. Once the changes had been made, the modifications were tested rigorously and, on March 20th, nearly one year to the day since the failed Falcon 1, a rebuilt Falcon 1 was ready to go again. It stood on the launch pad under a blazing yellow sun, the first new launch system in 30 years. Even before the launch, Musk had, in many ways, already made history. It took NASA months to turn around the Shuttle, yet Musk had brought a privately built rocket to the launch pad twice in one year. No one had ever done that.

The countdown reached the "one" mark but, except for orange flame, nothing happened; super-cold fuel had hit the engine at start-up, triggering an automatic abort. The solution was to drain back fuel on the first stage and reload fast with warm fuel – an exercise that took just 20 minutes. This time, the countdown proceeded through to launch and flames exploded from the rocket. Falcon 1 rose from the launch pad, accelerating quickly on a burst of flame. At 37,000 feet, it reached maximum dynamic pressure, known as Max Q – an event that was followed by first-stage separation. As the long second-stage nozzle glowed white hot, the team in El Segundo popped champagne corks. Three hundred seconds into flight, the rocket's second stage started to spin and wobble. Three minutes later, the stage was spinning faster. The second-stage engine flamed out, but the rocket continued to roll. At 11 minutes 11 seconds into flight, the video feed went blank.

Regardless, Musk claimed success with the launch of the vehicle. After all, he had gotten his rocket farther into space than any company ever had, with an engine designed entirely from scratch, all the major milestones were met, and the Defense Advanced Research Projects Agency (DARPA) reps had walked away happy, impressed at how quickly the SpaceX team had refueled the beefed-up Falcon 1.

By this time, Musk was breathing rarefied air, living in a 6,000-square-foot house (which had a domestic staff of five) in Bel Air hills, attending black-tie fundraisers, partying with Leonardo DiCaprio and Paris Hilton, and jetting to Richard Branson's private island on a private jet. He had also served as the inspiration for the Tony Stark character of *Iron Man* fame, as alluded to in this chapter's subtitle. When development of the first *Iron Man* movie was in its infancy, director Jon Favreau had a problem because he couldn't seem to make his main character, egotistical genius/superhero Tony Stark, come to life. In search of inspiration, Favreau turned to Robert Downey Jr, the actor hired to play Stark/Iron Man. Downey Jr recommended that the two should sit down with Musk and the rest, as they say, is history; the comparison was spot on. For those who are film buffs, the first *Iron Man* film briefly featured a Tesla Roadster parked in Starks's underground shop and, in the sequel, Musk made a cameo appearance with Stark, asking whether he might be able to design an electric jet. Musk was living a dream lifestyle, but it was also a very busy lifestyle. In between serving as a muse for an iconic comic book hero, Musk was also overseeing the progress of the development of Falcon 1, breaking ground at Launch Complex 40 at Cape Canaveral Air Force Station, monitoring the progress of the company's multi-engine test firing, slated for January 2008, and, of course, running Tesla.

Tesla hit a small hurdle in January 2008, when the company fired several key personnel to pare down costs after a performance review by Ze'ev Drori, Tesla's new CEO. The following month, a fifth round of funding added another US$40 million (Musk had contributed US$70 million of his own money at this time). Another hurdle of the personal variety struck Musk in September 2008 when his wife, Justine, announced she was divorcing him. They entered into counseling but, with Tesla and SpaceX to run, Musk had precious little time to solve his marital issues and, after three sessions, he filed for divorce. While he was dealing with the divorce, Musk also had his attention focused on an impending launch attempt of Falcon 1, which was being prepared at the US Army Kwajalein Atoll site in the Central Pacific. But the media weren't following the Falcon 1 story; thanks to Musk's wife being a prolific blogger, the public was treated to a blow-by-blow account of Musk's divorce proceedings. Predictably, the media followed the story and embellished and propagated it in a way only the news media know how. Some newspapers implied that Musk had run off with an actress, suggesting the reason for his divorce was his relationship with actress Talulah Riley. In reality, Musk had filed for divorce before he had met Riley. Other, equally creative news entities suggested that Matt Peterson, the president of Global Green and Justine's long-term boyfriend, played a part in Musk's decision to file for divorce. Again, the reality was markedly different: Peterson had been a mutual acquaintance of Musk and his wife for many years, and Musk's wife didn't enter into a relationship with him until after he had filed for divorce. The bottom line in the divorce was that there was no third party involved in the break-up at all. Still, it didn't stop CNBC from featuring the Musk divorce on *Divorce Wars*, a reality TV show that dissected the break-up piece by piece, explaining to those who are interested in messy divorces that Musk's wife was looking for US$6 million in cash and stock.

After Musk's divorce, and the successful launch of Falcon 1, a filing from the settlement was circulated, which stated Musk was broke and living on loans. The news created concern among Musk's critics, who suggested that since Musk couldn't manage his own money, he probably wasn't suited to run a multibillion-dollar car company. Like many venture capitalists, Musk's US$1.9 billion net worth was mostly tied up in his various ventures, including SpaceX. Like the erroneous divorce stories, the "broke millionaire" story was exactly that: erroneous. In reality, Musk had had to take a tough decision in late 2007 when Tesla was in a serious financial state and the only way out was for existing shareholders to recapitalize the company. Because the company was in such dire straits, there was no way to raise money externally and, rather than allowing Tesla to die, Musk committed almost all his cash reserves to the company, leaving a few million dollars to cover living expenses. Reality didn't satisfy the media though, who targeted his private jet as an extravagant indulgence and portrayed Musk as a playboy millionaire. Once again, the media accounts were way off the mark. To begin with, Musk didn't own any homes or expensive yachts, and hardly ever took vacations. He still doesn't. The guy is a self-confessed workaholic who has taken one holiday in the last four years. And the jet? In 2007, Musk made more than 200 business trips, spending 500 hours in the air. And that's not counting time spent in the airport! Try that crushing schedule without a jet.

By January 2009, Tesla, which had raised US$187 million and delivered 147 cars, attracted the interest of Germany's Daimler AG. On May 19th, 2009, the manufacturer of Mercedes acquired an equity stake of less than 10% of Tesla for a reported US$50 million. In July 2009, shortly after Tesla was approved to receive US$465 million in interest-bearing loans from the US Department of Energy, the company announced it had achieved corporate profitability for that month, having earned approximately US$1 million on revenue of US$20 million. More good news followed that same month after the second successful launch of Falcon 1, carrying the RazakSat satellite. Later that year, a new investor – Fjord Capital Partners, a specialized European private equity manager investing in the clean-energy sector globally – came on board and, early in 2010, Tesla Motors indicated its intention to file an initial public offering (IPO).

Shortly before its IPO offering, Musk hit the headlines again. This time, the subject wasn't divorce or SpaceX, but corporate excess. In a US Securities and Exchange filing, Tesla disclosed it had paid US$175,000 in fuel charges and landing fees associated with Musk's private jet. Now, US$175,000 is hardly an astronomical sum but, at a time when the public was particularly sensitive to corporate excess (remember, this was when certain motor companies were receiving government bail-outs), the news media asked why Musk couldn't have picked up the tab. There was no suggestion that Musk ran afoul of any laws because all the flights were for company business and, anyway, it's hardly unusual for CEOs to fly on private jets funded by their companies. For example, Apple paid US$16,000 for Steve Jobs's Gulfstream during the first nine months of 2009 and Google spent more than US$100,000 for aircraft-related costs for Google business in 2008. The difference, the media pointed out, was the discrepancy between Tesla, which lost US$31 million in the first nine months of 2009, Apple, which had earned a US$14 billion profit in 2009, and Google, which had made US$10 billion.

The US$175,000 that Musk had billed the money-losing company included trips to Washington, DC, to secure the project's US$465 million Department of Energy loan. You'll no doubt remember the rebuke from Congress that was provoked when the Big Three Detroit CEOs flew to Washington by private jet, asking for government help. As it geared up for an IPO, Tesla should have known that it would be subject to increased scrutiny and, given the motor industry controversy, perhaps Musk didn't anticipate how such flights might be perceived.

With the successful launch of Falcon 1, Musk could look to the future, part of which was to help humanity become a spacefaring civilization. To achieve that goal, he knew he had to establish a blueprint that followed a path of ever-increasing capability and reliability while simultaneously reducing costs – goals he had alluded to in his Senate Testimony in 2004. By following this plan, Musk hoped to ultimately reduce the cost of launching vehicles into orbit and increase the reliability of space access by a factor of 10. It was a bold ambition, but Musk believed that by eliminating the traditional layers of management internally and subcontractors externally, SpaceX could reduce its costs and accelerate decision-making and delivery. He also believed that by keeping most of the manufacturing in-house, SpaceX could reduce its costs, keep tighter control of quality, and ensure a tight feedback loop between the design and manufacturing teams. And, by concentrating on simple, proven designs with a primary focus on reliability, Musk was sure SpaceX could reduce the costs associated with complex systems operating at the margin. Coupled with the newly emerging market for private and commercial space transport, Musk hoped that his new model for reaching orbit would reignite humanity's efforts to explore and develop space. The following chapters outline this story.

2

The engine of competition

"With the advent of the ISS, there will exist for the first time a strong, identifiable market for 'routine' transportation service to and from LEO, and that this will be only the first step in what will be a huge opportunity for truly commercial space enterprise. We believe that when we engage the engine of competition, these services will be provided in a more cost-effective fashion than when the government has to do it."

Ex-NASA Administrator Dr Michael Griffin, November 2005

Following the spectacular success of SpaceX's Dragon flight to the International Space Station (ISS) in May 2012, it wasn't too surprising when the Obama Administration tried to take some of the credit. When Presidential Science Advisor John Holdren declared that the Obama Administration had made the Dragon flight possible, there were some rumblings among the Republicans, who felt justifiably aggrieved, since the Commercial Orbital Transportation Services (COTS) program had been proposed by the Bush Administration in 2005 and the COTS contract that funded the SpaceX mission had been awarded in 2006. On the subject of giving credit where credit's due, Michael Griffin, who led NASA during the Bush Administration, should also be mentioned, since it was he who conceived and funded the COTS federal seed money program that finally got the Dragon off the ground. However, the Obama Administration should still receive some credit because it was President Obama who increased commercial funding from US$500 million total to US$500 million a year and, as this book is being written, the Senate funding level for commercial spaceflight in fiscal 2013 stands at US$525 million (which was less than the US$836 million NASA had requested).

"I would like to start off by saying what a tremendous honor it has been to work with NASA. And to acknowledge the fact that we could not have started SpaceX, nor could we have reached this point without the help of NASA."

Elon Musk at a press conference after the launch
of the Dragon's first ISS flight

You may wonder why Musk thanked the space agency so profusely following the Dragon's historic flight to the ISS. After all, isn't this the same agency that many believe Musk is working to privatize out of existence? You may also wonder how much help NASA provided to SpaceX, since the public perception is that the commercial space race is funded by the companies themselves. In reality, the Dragon's mission was not a libertarian adventure. Far from it. It was, in fact, the result of a deeply collaborative effort between SpaceX and NASA. So, the purpose of this chapter is to provide an overview of the various funding initiatives that have helped SpaceX and the other companies in the commercial space race and also make sense of all those funding acronyms like SAA, COTS, CRS, CCDev, and CCiCap.

SPACE ACT AGREEMENTS (SAAs)

We'll start with the Space Act Agreements (SAAs). When it became clear during the Bush Administration that the Space Shuttle program was to be retired, NASA had to figure out how to transport cargo and humans to the ISS. Fortunately, the technology needed to launch into low Earth orbit (LEO) was firmly established and the private sector seemed to be ready to step up to the task. So, in 2006, NASA began investing private spaceflight companies through a program known as COTS. It wasn't the first time the government had given industry a helping hand; after all, it was the US government's early support of the railroad and the aviation industry that laid the foundation for private companies to succeed. Nevertheless, for the space industry, the agency's agreements with its newer partners such as SpaceX represented a sea change in the way it worked with the private sector. Rather than the traditional cost-plus model, in which companies were reimbursed the *cost* of a project *plus* an additional amount that guaranteed them a profit, SpaceX and its competitors would be working under SAAs, in which NASA paid increments of a fixed price once the companies accomplished previously agreed-upon milestones. In short, the companies would only get paid for what they achieved. If they didn't hit the milestones, they wouldn't get paid.

Another distinctive element of the SAAs is that they leave spacecraft design mostly in the hands of private companies. So, once a SAA is awarded, NASA doesn't tell the companies whether they are doing something right or wrong; it just lets them get on with the business of designing and building spacecraft. There is nothing stopping a company from asking NASA for advice (many of them do) but, at the end of the day, the companies don't build to any pre-established design specifications. For the most part, the companies are left to their own devices to figure out how to best achieve the agreed-upon milestones. It's an effective system because it's an approach that encourages innovation, reduces NASA's renowned tendency to micromanage, and, perhaps most importantly, it saves money. In fact, NASA reckoned SpaceX was able to build its Falcon 9 rocket for about one-third of what the agency would have spent on a similar project under its traditional model.

The SAAs have been a boon for SpaceX and the fledgling commercial spaceflight industry, since it is these agreements which NASA has used to funnel millions of dollars in investments into the companies. The agreements have been around for a while (NASA was first given the authority to enter into the agreements under the

Space Act of 1958) and there are a number of ways in which they can be used. Before the SAAs, the traditional government method of awarding a contract involved a request bid for a specifically outlined project, with cost overruns being paid by either the government or the contractor, according to the legal document. This method rarely worked well and, in the "learn-as-you-go" spaceflight arena, such a system has often been detrimental to the design and development of space systems and associated hardware. So, to harness collaboration between government, industry, and academic researchers, the SAAs were created. Today, there are three basic types of SAAs: reimbursable, non-reimbursable, and funded agreements. We'll take a look at each of these.

Under reimbursable agreements, a commercial firm or academic researchers may access "unique goods, services, or facilities" that NASA has but isn't fully using. In this type of SAA, a company might reimburse NASA for service use because, in return, it doesn't have to invest in those services alone. The non-reimbursable SAA is, as its name suggests, one in which no money is reimbursed. For example, recently, a university wanted to conduct tests on the effects of collisions of cometary matter. Under a non-reimbursable SAA, the university worked with Johnson Space Center (JSC) free of charge because NASA also needed a better way to analyze data from its Stardust and Deep Impact comet missions. For funded SAAs, NASA only pays if a company reaches each of several agreed-upon milestones (we'll discuss these in Chapters 6 and 7) in design, safety, and performance by a specific date. For example, before the Dragon flew to the ISS, SpaceX had to complete a design review of the vehicle and, by completing the review successfully, it received a US$75 million payday.

Since retiring the Shuttle, NASA has been using funded SAAs more and more to spur development of manned and unmanned spacecraft to realize its goal of finding a home-grown way to send crew and cargo to the ISS. In the longer term, NASA hopes the investment in innovation and infrastructure will foster a competitive market that can keep costs down for future missions in LEO and deep space. The bottom line is that these SAAs help NASA to keep costs down if project delays cause budget overruns because the extra expenses are borne by the firm, not the taxpayer. That's good reasoning for those who have to explain government costs to wallet-weary taxpayers and it's also good for NASA because the agency can defend the use of the agreements in an effort to stimulate innovation and growth in the commercial spaceflight industry.

Thanks to the SAAs, SpaceX in particular has been able to leapfrog ahead of where it would have been if it had tried to fly the Dragon (Figure 2.1) to the ISS in a purely commercial environment. But money is not the only reason SpaceX and the other private spaceflight companies are clamoring for government support. Entering into an SAA partnership with NASA confers a great source of legitimacy for an industry that, for some people, still smacks of science fiction. And that legitimacy is vital for attracting other customers as well as more traditional investors from Wall Street and investment banks. It doesn't hurt that NASA also helps private spaceflight companies navigate the complicated regulatory environment – assistance which will be especially important when passenger safety becomes an issue.

2.1 Dragon capsule. Courtesy: SpaceX

The SAAs also benefit the agency because, after 50 years of successful missions, LEO is no longer a real frontier but, by making transportation to and from the ISS the responsibility of private companies, NASA can be freed to develop the heavy lifters that will be needed to go to the asteroids and, hopefully, eventually to Mars. With the NASA budget expected to shrink over the next few years, the SAAs are a godsend for the agency because they will allow it to use what money it does get on science programs and exploratory missions.

> "For 50 years American industry has helped NASA push boundaries, enabling us to live, work and learn in the unique environment of microgravity and low Earth orbit. We're counting on the creativity of industry to provide the next generation of transportation to low Earth orbit and expand human presence, making space accessible and open for business."
>
> NASA's William Gerstenmaier

COMMERCIAL ORBITAL TRANSPORTATION SERVICES (COTS)

Announced on January 18th, 2006, COTS is a NASA program designed to coordinate the delivery of crew and cargo to the ISS by private companies. When the agency was forced to retire the Shuttle during the Bush Administration, NASA suggested that commercial cargo services to the ISS would be necessary through at least 2015 but, even with the successes of SpaceX, that date seems likely to slip to 2016 or later. But, thanks to various NASA funding programs such as COTS, by 2016 the agency may have a choice of systems to ferry cargo and astronauts to the ISS.

Since there are so many NASA funding programs, it's easy to get confused, so it's helpful to distinguish COTS from some of the other programs. To start with, it's important to distinguish COTS from the related Commercial Resupply Services (CRS) program: very simply, COTS relates to the development of the *vehicles*, whereas CRS deals with the *deliveries*. Also, whereas COTS involves a number of SAAs, with NASA providing milestone-based payments and binding contracts, CRS *does* involve legally binding contracts, which means the suppliers are liable if they fail to perform. Yet another program, the Commercial Crew Development (CCDev), discussed later in this chapter, is aimed just at developing crew rotation services. What all three programs do have in common is that they are all managed by NASA's Commercial Crew and Cargo Program Office (C3PO).

COTS isn't the first time NASA has explored programs for ISS services. In the mid-1990s, it funded Alternate Access, a preliminary study that convinced many entrepreneurs that the ISS would emerge as a significant market opportunity. While the ISS market never transpired, the retirement of the Shuttle forced NASA to revisit the purchase of commercial orbital transportation services on foreign spacecraft, since, at the time, NASA's own Crew Exploration Vehicle (CEV) wouldn't be ready until 2014. These foreign vehicles included the Russian Federal Space Agency's aging Soyuz (Figure 2.2) and Progress spacecraft, the European Space Agency's (ESA)

2.2 Designed for the Soviet space program in the 1960s, the venerable Soyuz capsule (originally built as part of the Soviet manned lunar program) is still in service today. Considered to be the world's safest, most cost-effective human spaceflight system, the Soyuz also happens to be very expensive, with the cost of a single seat priced at US$63 million. Courtesy: NASA

Automated Transfer Vehicle (ATV), and the Japan Aerospace Exploration Agency's (JAXA) H-II Transfer Vehicle (HTV). To bridge the gap to 2014, NASA initiated COTS, the first round of which took place in May 2006, when NASA selected six semi-finalist proposals for further evaluation. By the time the second-round selection took place on February 19th, 2008, NASA had made awards to Rocketplane Kistler (RpK), SpaceX, and the Orbital Sciences Corporation.[1]

---

[1]   RpK was originally awarded a contract worth US$207 million. The company received only US$32.1 million before NASA terminated their contract for failure to complete milestones. SpaceX was awarded US$278 million. In 2011, additional milestones brought the contract value to US$396 million. The Orbital Sciences Corporation was awarded US$170 million and, in 2011, additional milestones brought the contract value to US$288 million.

2.3 Known in early development as Taurus II, the Antares rocket is an expendable launch system being developed by the Orbital Sciences Corporation. It is designed to launch payloads weighing up to 5,000 kilograms into low Earth orbit. Courtesy: NASA

COMMERCIAL RESUPPLY SERVICES (CRS)

The development of the CRS program began in 2006 with the purpose of creating American commercially operated uncrewed cargo vehicles to service the ISS. Development of these cargo-carrying vehicles was under a fixed-price milestone-based program, which meant each company receiving funding had a list of milestones with a dollar value attached to them; companies only received the funding if they achieved the milestones. The first of these CRS contracts was awarded on December 23rd, 2008, when NASA awarded contracts to SpaceX and the Orbital Sciences Corporation (usually referred to as just "Orbital"). Under the contracts, SpaceX was to use its Falcon 9 rocket and the Dragon spacecraft to haul 20 tonnes of cargo to the ISS, while Orbital would use its Antares rocket (Figure 2.3) and Cygnus spacecraft. The contracts, worth a combined US$3.5 billion (US$1.6 billion to SpaceX and US$1.9 billion to Orbital) through 2016, were awarded based on the likelihood of rocket availability and the superior management structures and technical abilities demonstrated by the two companies' proposals.

When choosing among contenders, NASA faced a dilemma: whether to make just one contract award to the clear front-runner, SpaceX, or award a second contract to another finalist that happened to have the highest price and lowest score. After much deliberation, NASA decided it was important to the success of the ISS program to select multiple suppliers to maximize the probability of the station's resupply following the Shuttle's retirement. Ultimately, SpaceX was the clear winner because it offered the best technical proposal and the lowest overall price, but which company should NASA choose among the remaining contenders?

Those contenders were Orbital and PlanetSpace[2] of Chicago, and choosing between them required more deliberation. Orbital received the lowest score of the three finalists and charged the highest price, whereas the proposal from start-up PlanetSpace relied heavily on subcontractors Boeing, Lockheed Martin, and Alliant Techsystems (ATK), and didn't present a backup plan in the event that one of the subcontractors was unable to deliver. Eventually, NASA decided that Orbital's proposal was superior due to the serious management risks inherent in PlanetSpace's proposal and, because NASA had reservations about PlanetSpace's ability to successfully address the technical challenges of its proposal, the agency believed there was a low likelihood that the company could successfully perform the contract.

Another factor that affected the vote in Orbital's favor was the company's assurance that it could offer the full range of services by mid-2012. To sweeten the deal, Orbital offered a December 2010 "early bird" flight, which would double as a COTS demonstration flight. That assurance gave Orbital the edge despite its higher price and PlanetSpace was bumped.

COMMERCIAL CREW DEVELOPMENT (CCDev)

This is a multiphase space technology development program administered by NASA. Run by the agency's C3PO, the intent of the CCDev program is to stimulate development of privately operated crew vehicles to LEO. Under the program, at least two providers will be chosen to deliver crew to the ISS, hopefully no later than 2017. Unlike traditional space industry contractor funding used on the Space Shuttle, Apollo, Gemini, and Mercury programs, contract funding for the CCDev program contracts is explicitly designed to fund only specific subsystem technology development objectives that NASA wants for NASA purposes; all other system technology development is funded by the commercial contractor.

CCDev 1

In the program's first phase (CCDev 1), NASA provided US$50 million during 2010 to five companies, intended to foster research and development into human

---

[2]   PlanetSpace proposed using an existing rocket to provide initial cargo delivery capability in December 2011 before switching to the Athena 3 solid-fueled rocket Alliant Techsystems would build to support a full range of cargo services starting in late 2013.

spaceflight concepts and technologies in the private sector. Later that year, a second set of CCDev proposals was solicited by NASA for technology development project durations of up to 14 months. The proposals selected included Blue Origin, which was awarded US$3.7 million to develop an innovative "pusher" Launch Abort System (LAS) and composite pressure vessels. Boeing received US$18 million for development of its Crew Space Transportation (CST)-100 vehicle and Paragon Space Development Corporation was awarded US$1.4 million to develop an environmental-control and life-support system (ECLSS) air revitalization system (ARS) Engineering Development Unit designed to be used on different commercial crew vehicles. The Sierra Nevada Corporation (SNC) received US$20 million for development of Dream Chaser, its reusable spaceplane capable of transporting cargo and crew to LEO. The fifth company, United Launch Alliance (ULA), received US$6.7 million for an emergency distribution unit (EDU) for human-rating its evolved expendable launch vehicle (EELV) launch vehicles.

CCDev 2

On April 18th, 2011, NASA announced it would award up to nearly US$270 million to four companies as they met CCDev 2 objectives. These objectives included the capability of a vehicle to deliver and return four crewmembers and their equipment, provide crew return in the event of an emergency, serve as a 24-hour safe haven in the event of an emergency, and remain docked to the ISS for 210 days. Winners of funding in the second round of the CCDev program included Blue Origin, which was awarded US$22 million to develop advanced technologies in support of its orbital vehicle, including launch abort systems and restartable hydrolox (liquid-hydrogen/liquid-oxygen) engines, and SNC received US$80 million to develop phase 2 extensions of its lifting body-inspired Dream Chaser (Figure 2.4) spaceplane.

SpaceX was awarded US$75 million to develop an integrated launch abort system design for its Dragon spacecraft. The system, reputed to have advantages over the more traditional tractor tower approaches used on prior manned space capsules, would be part of the company's Draco maneuvering system, currently used on the Dragon capsule for in-orbit maneuvering and de-orbit burns. Industry juggernaut Boeing proposed additional development for their seven-person CST-100 spacecraft (Figure 2.5), beyond the objectives for the US$18 million received from NASA in CCDev 1. Designed to be used up to 10 times, the capsule would have personnel and cargo configurations, and is designed to be launched by different rockets.

While not selected for funding, ULA's proposed development work to human-rate the Atlas V rocket persuaded NASA to enter into an unfunded SAA with the company. A similar agreement was made with ATK and Astrium, which had proposed development of the Liberty rocket derived from Ares I and Ariane 5. A third unfunded agreement was made between the agency and Excalibur Almaz Inc. (EAI), which had proposed developing a crewed system incorporating modernized Soviet-era space hardware designs intended for tourism flights to orbit; EAI's

2.4 Dream Chaser. Courtesy: NASA

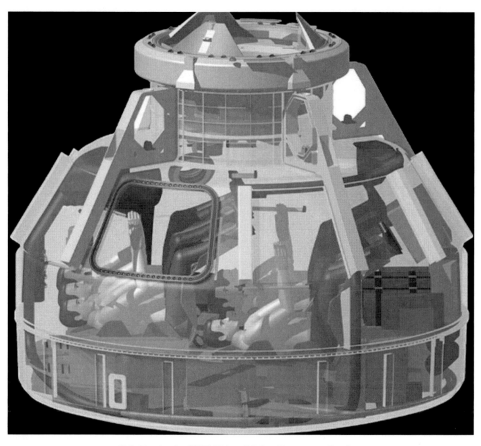

2.5 Boeing's CST-100 vehicle. Courtesy: Boeing

concept for commercial crew to the ISS was to use the company's three-person space vehicle with an intermediate stage and fly the integrated vehicle on a commercially available launch vehicle.

In common with many government-funding processes, to the outsider, NASA's rationale for CCDev awards seemed difficult to grasp. Consider the following proposals. One company proposed continuing work on a project that NASA had already funded to human-rate a pair of highly reliable rockets that at least three companies wanted to use to launch their commercial spacecraft. Another company requested funds to build a new booster that had never flown and which no one intended to use! Which one did NASA fund? Neither. Puzzled? Welcome to the perplexing world of NASA contract awards. The aforementioned companies were ATK, which proposed its Liberty rocket, and ULA, which is developing technologies to human-rate its Atlas V and Delta IV launchers. Why didn't they receive funding? Surprisingly, at least part of the reason seemed to have had little to do with the quality of the proposals. The rationale went something like this: spacecraft proposals were weighted higher than those for launch vehicles for the simple reason that

American companies have considerable experience developing launch vehicles but no US companies have successfully developed a crew-carrying spacecraft – at least not in the last 30 years. Given this emphasis, it is easier to understand why Boeing and SNC received funds to develop their human vehicles and why SpaceX received funding to human-rate its Dragon spacecraft and Falcon 9 rocket, the development of which NASA has been funding under the COTS program. Blue Origin is also developing a human-rated vehicle, so it received money to develop its biconic capsule and reusable rocket. Meanwhile, proposals that focused solely on rocket development received nothing.

While the experience in developing launch vehicles versus the experience in developing spacecraft provided part of the justification when it came to awarding funding, the strength and weaknesses of each program also factored into the decision-making, so it's worthwhile taking a look at how NASA makes these assessments. Two factors NASA is particularly interested in are the company's Technical Approach and Business Information. These factors are color-coded, with green indicating a High Level of Confidence and white indicating a Moderate Level of Confidence. For example, ULA, which had sought US$40 million in CCDev 2 funding, rated surprisingly low. NASA's reviewers identified several strengths in ULA's proposal, among them the company's use of existing flight-proven vehicles and infrastructure, its adaptable emergency detection system, a strong performance capability for crew-abort scenarios, and an effective and integrated organizational structure. Its business information didn't fare too badly either, since it was deemed suitable to deliver proposed capabilities, it had a strong, highly experienced management team, possessed the requisite facilities, and had experienced and knowledgeable suppliers. With such a strong résumé, you may be wondering why the company wasn't funded, but the reviewers also found weaknesses, among them a lack of definition of a critical path to an initial launch capability and correlation to CCDev 2 efforts, and a failure to adequately describe the commercial market to which it would provide products and services.

Ultimately, NASA deemed that ULA's work on their existing launch vehicles was not on the critical path for any crew-transportation system and therefore the company did not accelerate the availability of crew-transportation capabilities, which, after all, was a primary goal of the funding announcement. Nevertheless, while this assessment was true, it seemed strange that NASA would deny funding to a company that is developing a rocket that two CCDev 2-funded companies hoped to use: SNC and Blue Origin, which both received CCDev 2 funding, had stated that they wanted to use ULA's Atlas V (Figure 2.6) rocket to launch their crew vehicles (Blue Origin eventually shifted to its own reusable launcher). Boeing, which also received CCDev 2 funds, had also expressed its intent of using Atlas V to launch its CST-100 spacecraft, although the vehicle is being designed for multiple launchers.

At first glance, another strange case was the non-funding of ATK's Liberty rocket. This launcher comprises a first stage derived from the canceled Shuttle-derived Ares I booster and the second stage of Europe's Ariane 5. From the NASA reviewer's perspective, while these technologies had excellent flight heritages, the problem was that no one had committed to flying on the rocket. You can understand

2.6 United Launch Alliance's Atlas V rocket. Courtesy: ULA

2.7 Liberty rocket. Courtesy: NASA

the agency's concern: NASA could fund the Liberty (Figure 2.7) all the way through the development phase but there would always be the risk that no spacecraft developer would select the launch vehicle as part of its design!

Another black mark against ATK was their failure to provide NASA with sufficient details to assess launch vehicle environments (such as staging and abort scenarios) on the company's proposed upper stage or at the crewed spacecraft interface. While the company provided a solid technical approach, their details on environments didn't provide NASA with enough confidence in accelerating this launch vehicle for use with the variety of different crew spacecraft. So, rather than using limited CCDev 2 funds on a launch vehicle with a questionable technical approach, the agency instead decided to select an extra spacecraft.

COMMERCIAL CREW INTEGRATED CAPABILITY (CCiCap)

On August 3rd, 2012, NASA announced awards made to three American commercial companies under yet another funding program – CCiCap. Advances made by these companies under the signed SAAs through the agency's CCiCap initiative are intended to lead to the availability of commercial human spaceflight services for government and commercial customers., The CCiCap partners were SNC, which received US$212.5 million, SpaceX, which received US$440 million, and the Boeing Company, which received US$460 million.

As an initiative of NASA's Commercial Crew Program (CCP), CCiCap is an administration priority. The objective of the CCP is to facilitate the development of a US commercial crew space transportation capability with the goal of achieving safe, reliable, and cost-effective access to and from the ISS and LEO. After the capability is matured and expected to be available to the government and other customers, NASA plans to contract to purchase commercial services to meet its station crew-transportation needs.

The new CCiCAP agreements follow two previous initiatives by NASA to spur the development of transportation subsystems and represent the next phase of US commercial human space transportation, in which industry partners develop crew-transportation capabilities as fully integrated systems. Between now and May 31st, 2014, NASA's partners will perform tests and mature integrated designs. This will then set the stage for a future activity that will launch crewed orbital demonstration missions to LEO by the end of the decade.

3

The engines: The workhorses of commercial spaceflight

In many commercial ventures, it isn't unusual to develop a product without developing an engine to power it. This is because there is a ready availability of engines in different power ranges, which allows companies to produce an array of machines almost as quickly as they can be imagined. Even in highly specialized applications, where the final product is built in limited production numbers, the power plant is usually common almost to the point of being mundane. This is not the case when it comes to space launch systems.

Until quite recently, the development of launch vehicles was characterized by vehicle-specific engines which consumed a significant part of the rocket's development budget. But, in the last decade or so, rocket designers have begun to source common off-the-shelf solutions. For example, when Lockheed Martin was looking for an engine to power its Atlas 5 booster, it decided to use the Russian RD-180 (Figure 3.1).

Another example is the Orbital Sciences Corporation (usually referred to as just "Orbital"), which decided to use refurbished Russian NK-33 engines to boost its Taurus 2 launch vehicle (Figure 3.2). The choice of the NK-33 was a surprise to many in the commercial spaceflight industry, since these engines had a less-than-reliable reputation, having been the engine of choice for the now bankrupt Kistler K-1 and also the failed Soviet N-1 booster program whose termination resulted in these engines being sequestered by Orbital.

And then there is the case of SpaceX, which could have gone the same route as Orbital or Lockheed Martin, but instead chose to develop their engine in-house, beginning with the Merlin.

MERLIN

With 512,000 newtons (115,000 pounds of thrust) of sea-level thrust, the diminutive Merlin (Figure 3.3) is perfect for second- or orbital-stage applications. It's a relatively simple engine by design that runs on liquid oxygen (LOX) and rocket-

3.1 RD-180 rocket engines. Courtesy: *www.thehalsreport.com*

3.2 Artist's conception of the Orbital Sciences Corporation's Taurus II rocket. Courtesy: *www.parabolicarc.com*

3.3 SpaceX's Merlin engine. Courtesy: SpaceX

grade kerosene in a combustion process that utilizes a straightforward open-cycle process as opposed to staged.[1]

The rocket was designed by Tom Mueller, who brought his TRW experience to bear on the nation's first new large liquid-fuel rocket engine in 40 years. Unlike more complex engines that mix fuel and oxidizers at multiple points, the Merlin uses a

---

[1]   In an open-cycle process, some of the propellant is burned in a gas generator and the resulting hot gas is used to power the engine's pumps. The gas is then exhausted, hence the term "open cycle". In the staged combustion cycle, some of the propellant is burned in a pre-burner and the resulting hot gas is used to power the engine's turbines and pumps. The exhausted gas is then injected into the main combustion chamber, along with the rest of the propellant, and combustion is completed.

single injector. This injector, which draws upon a long heritage of space-proven engines, is the heart of the Merlin engine. The injector is of the pintle type that was first used in the Apollo Program for the Lunar Module Landing Engine (LMDE) that powered the Lunar Excursion Module (LEM). Interestingly, the company that built the LEM descent engine was TRW.

If you're wondering how a pintle injector works, think about your garden hose for a moment. At the end of the hose is a nozzle. Turn it one way and you get a steady, narrow stream of water shooting out in a long arc; turn it the other way to shut it off and you get a cone-shaped fan spray. When you look into the nozzle, you'll see a round pintle that moves back and forth as you turn the outer casing one way and the other. The fan-shaped water spray is what the fuel and oxidizer spray looks like inside the rocket engine, the only difference being that the spray of water doesn't combust!

In 2000, TRW demonstrated a newer design of an engine that used a pintle injector – the TR-106, also called the Low Cost Pintle Engine (LCPE). The LCPE generated 650,000 pounds of thrust, which was more than the 400,000 pounds of thrust generated by the Space Shuttle Main Engine (SSME). The LCPE, which was one of the largest liquid-rocket engines built since Saturn F-1 engines powered Apollo flights in the 1970s, was designed as a simple, easy-to-manufacture, low-cost engine constructed from common steel alloys using standard industrial fabrication techniques. Instead of using expensive regenerative cooling, the LCPE employed ablative cooling techniques and featured the least complex type of rocket propellant injector – a single-element coaxial pintle injector.

The LCPE's design was borne of TRW's goal to design an engine that minimized cost while retaining excellent performance rather than following the tradition of rocket design that sought maximum performance and minimum weight; by doing this, TRW hoped to reduce the cost of launch vehicles and enable access to space for government and commercial customers.

The LCPE was subjected to hot-fire testing at 100% of its rated thrust as well as at a 65% throttle condition at NASA's John C. Stennis Space Center (SSC) in Mississippi. In between tests, TRW changed the pintle injector configuration three times to investigate the engine's performance envelope. Throughout the tests, the engine demonstrated rock-solid performance, stability, and versatility. Since the engine provided the inspiration for the Merlin design, it's worth taking a look at some of the LCPE's features.

Perhaps one of the engine's signature features was its scalability. The LCPE was scalable over a range of thrust levels and propellant combinations, enabling it to be easily adapted to a number of launch vehicles capable of lifting anything from 200 to 200,000 pounds to low Earth orbit (LEO). The LCPE could also power the first stage of an evolved expendable launch vehicle (EELV) multistage launch vehicle and scaled-down versions of the engine could be used for the vehicle's second stage.

Another of the LCPE's positive traits was combustion stability, meaning it could operate over a wide range of operating conditions thanks to the unique injection and combustion flow fields created by the pintle injector. It was the pintle injector that was the key to many of the LCPE's performance attributes. Simple in design, the

3.4 One of the new Merlin 1D engines that will power the Falcon 9 rocket on future missions is test fired at SpaceX's Rocket Development Facility in McGregor. Courtesy: SpaceX

pintle injector contained only five parts yet it was this element that permitted the LCPE's deep throttle capability.

Given the LCPE's sterling performance and the huge strides TRW had taken towards providing more affordable access to space, it was a surprise to many that NASA canceled further work on the engine. Less surprising was Tom Mueller's decision to go with a single-pintle design for SpaceX's Merlin rocket engines.

Merlin's design was driven by the Falcon 1 rocket. The goal was to design an engine that could withstand a 160-second burn (Figure 3.4), which would be long enough for the first stage of Falcon 1 to reach an altitude of 90 kilometers. At this altitude, the second stage's smaller engine would kick in to boost the cargo to 130 kilometers and LEO.

Designing a rocket engine from scratch has never been an easy task but in October 2003, Mueller and his engineers figured they were ready to fire up a Merlin on a test stand. In this first Merlin variant, propellant was fed via a single-shaft, dual-impeller turbo-pump operating on a gas generator cycle. The turbo-pump also provided the high-pressure kerosene for the hydraulic actuators, which recycled into the low-pressure inlet – a design feature that eliminated the need for a separate hydraulic power system. This design also meant that thrust vector control failure caused by running out of hydraulic fluid wasn't possible. Another use of the turbo-pump was providing roll control by actuating the turbine exhaust nozzle. Combining the three functions into one device allowed engineers to verify all systems were functioning before the vehicle was allowed to lift off and also resulted in a significant improvement in system-level reliability. Unfortunately, the test didn't go too well. During the run, the Merlin's exhaust began to melt the metal in the engine's throat and the intense heat endangered seals responsible for governing propellant. After 60 seconds, the engineers decided to shut it down – any longer and the engine might have blown up.

Over the next 15 months, Mueller and his team of engineers went to work to troubleshoot the bugs. One of the fixes was reducing the amount of LOX entering the injector – a solution that made the engine run cooler and at lower power. Another fix was strengthening the seals. To prevent more heat damage, Mueller treated the Merlin's nozzles with resin impregnated with silicon fibers – an ablative coating designed to char and flake off as the engine ran, taking damaging heat with it.

Once the bugs were fixed, the engineers once again took their place in the bunker for the second test, which was planned to be a full-mission duty cycle, equating to the time needed to deliver a payload to LEO. The test proceeded without a hitch, with the Merlin shutting down after 162.2 seconds. It was an impressive performance, not just in terms of how well the Merlin performed, but also because the Merlin was the first rocket engine to be developed in the US since Rocketdyne's RS-68 engine (for Boeing's Delta IV) in the early 1990s and only the second since the late 1970s when Rocketdyne developed the SSME. Now all Mueller's team had to do was mount the engine on the Falcon 1 rocket and launch it.

As described in Chapter 1, the rocket was doomed before ignition due to the salty Pacific air that corroded an aluminum nut on the engine, causing a leak. The result was catastrophic. The spilled fuel caught fire and, 34 seconds after launch, flames burned through a pneumatic line and shut down the engine, causing the rocket to crash into the Pacific seconds later. A year later, sloshing fuel in the second stage of another Falcon 1 caused the rocket to spin out of control before reaching orbit. Then, during the third flight in August 2008, the first stage collided with the second stage shortly after separation.

Mueller's team persevered and, less than two months later, another Falcon 1 roared from its South Pacific launch pad. After a week spent reviewing data, Mueller confirmed the flight had gone really well. It was the culmination of six years of hard work by a very talented team and a relief to Musk who, two weeks before the flight, had received the AIAA George M. Low Space Transportation Award for the most outstanding contribution to the field of space transportation.[2]

Falcon 1's fourth flight achieved orbit, with the first burn terminating at a 330.5-

---

[2] Established in 1988, the AIAA's George M. Low Space Transportation Award honors the achievements in space transportation by Dr George M. Low, who played a leading role in planning and executing the Apollo missions. The biennial award is presented for a timely outstanding contribution to the field of space transportation. Musk's citation read: For outstanding contribution to the development of commercial space transportation systems using innovative low-cost approaches.

3.5 Kestrel engine being test fired. Courtesy: SpaceX

kilometer altitude and 8.99° inclination, which was pretty damn close to the intended insertion of 330-kilometer altitude and 9.0° inclination. The second burn tested the restart capability, allowing the upper stage to coast for 43 minutes, boosting the orbit to 621 kilometers.

KESTREL

The Merlin wasn't the only rocket engine that SpaceX built around the pintle architecture. The Kestrel (Figure 3.5) is a high-efficiency, low-pressure vacuum engine that features a vacuum thrust of 6,245 pounds of thrust and a vacuum-specific impulse of 325 seconds. The Kestrel, which was developed by SpaceX to propel the upper stage of the Falcon 1 rocket, doesn't have a turbo-pump and is a pressure-fed rocket. It is ablatively cooled in the chamber and throat and radiatively cooled in the niobium nozzle, which is designed to be super strong at extreme temperatures and is highly resilient to impact. Thrust vector control is achieved by electromechanical actuators on the engine dome for pitch and yaw, while roll and attitude control during the coast phases is provided by helium cold gas thrusters.

A multiple-restart[3] capability on the upper stage is provided by a highly reliable triethylaluminum–triethylborane (TEA–TEB) pyrophoric system. TEA is a volatile, highly pyrophoric compound that ignites immediately upon exposure to air.

---

[3]   In a multi-manifested mission, a restart capability allows delivery of separate payloads to different altitudes and inclinations.

Industrially, the compound is used as an intermediate in the production of fatty alcohols, which are then converted to detergents. Since TEA is one of the few substances pyrophoric enough to ignite on contact with cryogenic liquid oxygen (LOX), the substance is especially desirable as a rocket-engine igniter. TEB, TEA's cousin, which is also strongly pyrophoric, has a long history in the aerospace world, having been used to ignite the JP-7 fuel in the Pratt & Whitney J58 engines that powered the SR-71 Blackbird spy-plane.

MERLIN VARIANTS

For those who have followed the SpaceX story, the Merlin's development can be a little confusing because there have been a number of variants, some of which have flown and some of which haven't. What follows is an account of the variants in the SpaceX rocket engine inventory.

Merlin 1A

The initial version, Merlin 1A, used an expendable, ablatively cooled carbon-fiber composite nozzle and produced 340 kilonewtons (77,000 pounds of thrust) of thrust. This variant flew only two times, first on March 24th, 2006, when the engine caught fire and failed due to a fuel leak shortly after launch, and the second time on March 21st, 2007, when it performed successfully.

Merlin 1B

The Merlin 1B rocket engine was an upgraded version of Merlin 1A. Capable of producing 380 kilonewtons (85,000 pounds of thrust) of thrust thanks to a turbine upgrade, the initial use of Merlin 1B was to be on the Falcon 9 launch vehicle, which would have used a cluster of nine (hence the designation) Merlin 1Bs on the first stage. However, due to experience from Falcon 1's first flight, Merlin 1B was never used on a flight vehicle and SpaceX moved its Merlin development to Merlin 1C.

Merlin 1C

Merlin 1C uses a regeneratively cooled[4] nozzle and combustion chamber. It was this

---

[4] As you can imagine, firing a rocket results in tremendous heat, with combustion temperatures reaching 2,500–3,600 K. To cool the thrust chamber, rocket designers use a variety of chamber-cooling techniques, the most widely used of which is regenerative cooling. This method cools the thrust chamber by flowing high-velocity coolant over the back side of the chamber hot gas wall to convectively cool the hot gas liner. The coolant with the heat input from cooling the liner is then discharged into the injector and utilized as a propellant.

engine that powered the successful fourth Falcon 1 flight in September 2008 (it also powered the third unsuccessful flight) and Falcon 9 on its maiden flight in June 2010. When configured for Falcon 1 vehicles, Merlin 1C had a sea-level thrust of 350 kilonewtons (78,000 pounds of thrust) and a vacuum-specific impulse of 304 seconds but, when configured for Falcon 9, the sea-level thrust is 560 kilonewtons and the specific impulse is 300 seconds.

Merlin Vacuum (1C)

The follow-up to Merlin 1C was the Merlin Vacuum, which featured a larger exhaust section and a significantly larger expansion nozzle to maximize the engine's efficiency in the vacuum of space. In common with its predecessor, the Merlin Vacuum's combustion chamber is regeneratively cooled while the niobium alloy expansion nozzle is radiatively cooled. The engine produces a vacuum thrust of 411 kilonewtons and a vacuum-specific impulse of 342 seconds. On January 2nd, 2010, the first-production Merlin Vacuum engine underwent a full-duration orbital insertion firing lasting 329 seconds before being flown on the second stage for the inaugural Falcon 9 flight on June 4th, 2010.

An unplanned test of a modified Merlin Vacuum engine took place in December 2010. Shortly before the second flight of Falcon 9, two cracks were discovered in the engine's niobium alloy sheet nozzle. Engineers decided to cut off the lower 1.2 meters of the nozzle and launch two days later, since the extra performance that would have been gained from the longer nozzle wasn't necessary to meet the mission objectives (even with the shortened nozzle, the engine still placed the second stage into an orbit of 11,000 kilometers).

Merlin 1D

In June 2012, SpaceX made another addition to its stable of rocket engines by successfully test firing the Merlin 1D engine. Designed to propel future SpaceX vehicles into the thermosphere, Merlin 1D is the follow-up to the Merlin engines used to ferry Dragon to the International Space Station (ISS) the previous month. Thanks to an improved thrust-to-weight ratio, Merlin 1D is the most efficient booster engine ever built, featuring a vacuum thrust of 690 kilonewtons (155,000 pounds of thrust) and a vacuum-specific impulse of 310 seconds. In addition to sporting impressive propulsion numbers, the new Merlin also features a throttle capability from 100% to 70% and a 160:1 thrust-to-weight ratio while still maintaining the structural and thermal safety margins (the highest ever achieved for a rocket engine) needed to carry astronauts. The engine achieved its first full-mission-duration firing in June 2012, in a test that featured multiple restarts at target thrust and specific impulse. The engine firing was for 185 seconds with 147,000 pounds of thrust – the duration and power required for a Falcon 9 rocket launch. At the time of writing, SpaceX plans to have the engine ready for the sixth flight of Falcon 9 in 2013. If you're the sort of space fanatic who enjoys watching big flames being pushed into a concrete chamber, you'll most likely enjoy watching the

3.6 SpaceX's McGregor test stand. Courtesy: SpaceX

YouTube video of the June 2012 test, which took place at SpaceX's rocket development facility (Figure 3.6) in McGregor, Texas (see *www.youtube.com/ watch?v=976LHTpnZkY*).

There's little doubt the Merlin 1D engines will make the next generation of Falcon rockets a lot more powerful but, on the scale of space exploration, where do these engines fit? Well, nine Merlin 1C engines, the standard configuration for Falcon 9's first stage, generate about 1.1 million pounds of thrust at launch, whereas the Merlin 1D engine 185-second test fire in June 2012 delivered 147,000 pounds of thrust, so nine of these new engines clustered on Falcon 9's first stage will give the rocket nearly 1.5 million pounds of thrust at launch. Applied to the Falcon Heavy, which

3.7 Engineers at the Marshall Space Flight Center install the F-1 engines on the S-IC stage thrust structure at the S-IC static test stand. Courtesy: NASA

uses 27 engines in its first stage, these upgraded Merlin engines will give the rocket an impressive 3.8 million pounds of thrust at launch. Now, let's compare these impressive numbers with the Saturn F-1 (Figure 3.7), which was capable of generating 1.5 million pounds of thrust.

Developed by Rocketdyne, the F-1 (five of them) rocket engine was used in the S-IC first stage of each Saturn V, which served as the main launch vehicle in the Apollo program. It is still the most powerful single-chamber liquid-fueled rocket engine ever developed. During launch, this behemoth burned 1,789 kilograms of LOX and 788 kilograms of RP-1 *every second*, generating 6.7 million newtons of thrust. During their two and a half minutes of operation, the cluster of F-1s propelled the Saturn V vehicle to a 68-kilometer altitude and a speed of 9,920 kilometers per hour. Each F-1 engine had more thrust than three SSMEs (no slouch in the propulsion department) combined. Despite being based on 45-year-old technology, the F-1 engines haven't been forgotten; in 2012, it was proposed to use the engines to launch the Space Launch System (SLS) rocket that NASA hopes will eventually send its astronauts to the asteroids and perhaps Mars.

So, while the Merlin 1D engine doesn't pack the punch of the F-1, the June 2012 firing showed the company was on track for bringing the company a step closer to the rocket power it needs in the future.

Merlin 2

The Merlin-engine iteration doesn't stop at Merlin 1D. At the AIAA Joint Propulsion conference on July 30th, 2010, McGregor rocket development facility director Tom Markusic shared some information with the media about the development of a new SpaceX engine: Merlin 2. Merlin 2, which is being designed to launch conceptual super-heavy-lift launch vehicles, is a LOX/RP-1-fueled engine, capable of a projected 7,600 kilonewtons of thrust (1,700,000 pounds of thrust) at sea level and 8,500 kilonewtons (1,920,000 pounds of thrust) in a vacuum. According to the presentation at AIAA, SpaceX reckons the engine could be qualified in three years for US$1 billion.

DRACO AND SUPERDRACO

On a smaller scale than the Merlin family of rocket engines is the Draco, a small hypergolic rocket engine/thruster designed for use on Dragon and the upper stage of the Falcon 9 rocket.

The Draco thrusters, 18 of which are used on Dragon (four are used on the Falcon second stage), generate 400 newtons (90 pounds-force) of thrust using a mixture of monomethyl hydrazine fuel and nitrogen tetroxide oxidizer. Used for attitude control and maneuvering, the thrusters are dual-redundant in all axes, which means any two can fail and astronauts will still have complete vehicle control in pitch, yaw, roll, and translation.

An addition to the Draco family was announced by SpaceX on February 1st, 2012, when the company told the media it had completed the development of a more powerful version of the Draco thruster called, appropriately enough, the Super-Draco (Figure 3.8). The SuperDraco, a throttleable engine with multiple-restart capability, is designed for SpaceX's launch abort system (LAS) on the Dragon spacecraft. In the event of a launch abort, eight SuperDracos will fire for five seconds at full thrust. The engine, the development of which is partially funded by NASA's CCDev 2 program, generates a thrust of 67,000 newtons (15,000 pounds of thrust), making it the second most powerful engine developed by SpaceX, more than 200 times more powerful than the Draco (and twice as powerful as the Kestrel) with about one-ninth of the thrust of a Merlin 1D engine. In addition to using the SuperDraco thrusters for the LAS, SpaceX also plans to use them for powered landings on Earth and, one day, the SuperDracos may even be used as retro-propulsion to decelerate a manned Mars lander.

RAPTOR

The Raptor liquid-hydrogen/LOX second stage was first mentioned by Max Vozoff at the AIAA Commercial Crew/Cargo Symposium in 2009. Few details have been publicly released, but the engine is in active development and it is believed it might be

3.8 SuperDraco being test fired. Courtesy: SpaceX

used for high-energy destinations like geostationary orbits, Mars, or perhaps asteroid missions. The engine would be powered by LOX and liquid hydrogen – a design that would enable a greater mass to be boosted into orbit. If the Raptor becomes operational, it might replace the Merlin Vacuum engine in the current Falcon 9 on high-performance launches.

FUTURE PROPULSION SYSTEMS

Over the past few years, SpaceX has deliberated about the usefulness of nuclear-powered spacecraft. Since the company wants to send its rockets to Mars, some SpaceX engineers believe a nuclear-powered upper stage might be one way to get to the Red Planet. In the mission architecture SpaceX proposes, a beefed-up version of the Falcon 9 rocket would be used to lift cargo, spacecraft, and crew into orbit. Once there, SpaceX wants government-built nuclear-powered rockets to take over. Of course, there have been several manned Mars-mission proposals over the years that have included the use of NERVA (Nuclear Engine for Rocket Vehicle Application), a nuclear thermal engine developed by NASA in the 1960s. The goal of the NERVA project was to use a nuclear reactor as a rocket motor. The project made good progress, with a number of test reactors being built, before President Nixon killed the program in 1973 over cost concerns. Before its demise, the project had shown that nuclear rockets were a far more fuel-efficient way to power spacecraft than conventional rockets (fissile uranium has an energy density close to 80,000,000 megajoules per kilogram, while kerosene is more like 40 megajoules per kilogram) although, in common with conventional rockets, they weren't foolproof. One big headache was that all the super-heated hydrogen would corrode the fuel cladding, causing the engine to leak radioactive uranium. But, while this is an issue on Earth, it's less of a concern in orbit, so using a NERVA derivative to get to Mars just might be an idea worth pursuing. Of course, there's still the financial challenge, and it's worth remembering that SpaceX prides itself on being cheaper and just as good when it comes to conventional spaceflight, so the fact they're willing to cede nuclear propulsion to the government gives you some idea of just how difficult nuclear propulsion can be.

With an energy density a million times greater than conventional rocket fuels, the attractiveness of going nuclear is undeniable but, even assuming the engineering problems with a nuclear-powered heavy-lift vehicle can be resolved, there are still sizable obstacles in the path of a hypothetical SpaceX–NASA hybridized nuclear Mars rocket. One of the more challenging hurdles is the 1972 Liability Convention, which affects the planning and execution of almost all space launches but has additional requirements for vehicles carrying nuclear material. Then there is the 1992 UN Resolution on Nuclear Power Sources, which is pretty clear in laying out the guidelines and criteria for nuclear safety aboard spacecraft using nuclear material for non-propulsive purposes. Let's take a look at each of these.

First is the 1972 Liability Convention, also known as the Convention on International Liability for Damage Caused by Space Objects. This Convention is one of the Big Four space-law treaties and it's been ratified by 90 countries, including the US, China, and Russia. In layman terms, the Convention holds launching states liable for damage to other nations caused by the launching state's space object falling to Earth. For example, let's say that SpaceX launched a private commercial satellite on a Falcon 1 and the satellite malfunctioned and ended up crashing in downtown London, England. In this scenario, the US would be responsible for paying for the damage to the British government, including any loss of life, personal injury, and damage to private and public property. In most cases, the US taxpayer would probably avoid having to pay for the entire cost of the damage, since US law requires launch services providers such as SpaceX to carry insurance covering precisely these kinds of accidents.

Now, imagine a scenario that involved the launching of a nuclear-powered spacecraft. If SpaceX was the launch provider, their legal team would face some unique challenges in determining the scope of potential damage, due to the lasting effects of radioactive contamination in the event of a catastrophe. Obviously, they would attempt to reduce risk by determining the safest flight profile to minimize potential damage to nations and people and implement safety measures such as installing ruggedized impact shells (this measure was implemented on the Cassini mission[5]). Still, even with all these measures, it would be a tough sell to the public.

Next is the 1992 UN Nuclear Power Principles Resolution, which, while not as

---

[5]   Cassini carried 33 kilograms of plutonium spread among 18 modular units, each with its own heat shield and impact shell. These containers were tested to destruction and found to release plutonium only in much harsher conditions than were likely to be present during launch.

tough as the Liability Convention (UN Resolutions are not explicitly intended to be binding on member nations), is still part of customary international law. While this gentleman's agreement between nations asks that countries follow rules without explicitly drafting up a document saying they are bound to follow them, customary international law usually carries some weight in courts around the world. The Nuclear Power Principles Resolution lays out guidance for the use of nuclear power sources in space. Interestingly, since the Resolution only lays down operating principles for *non-propulsive* uses of nuclear power sources, it would arguably apply to spacecraft like Cassini, but not to NERVA-propelled rockets! We'll discuss the nuclear option again in Chapter 8 but, for now, let's return to the Merlin.

MERLIN LEGACY

Why has the Merlin design been so successful? Well, the engine is a manageable size and, as far as rocket designs go, the system is relatively basic, since the Merlin runs on arguably the simplest combination of fuels – LOX and rocket-grade kerosene – and utilizes a fairly straightforward open-cycle (as opposed to staged) combustion process. Compared with liquid-hydrogen-based systems, RP-1 offers an advantage in terms of ease of system design and operational handling. It is this simplicity that confers a reduced cost and complexity in terms of propellant piping, seals, valves, and insulation, as well as the ability to incorporate smaller fuel tanks.

By selecting just one basic engine to be used for its family of launch vehicles (with the exception of the Falcon 1 second-stage Kestrel engine), SpaceX has been able to not only reduce design and production costs, but also allow engineers to acquire a rapid buildup in experience (don't forget, 10 engines are utilized in each Falcon 9 flight). Assuming that SpaceX's current manifest proceeds, the company might fly in excess of 200 Merlin engines in the next five years leading up to 2017 – plenty of time to examine and refine engines post flight with an eye towards future improvements. With that depth of potential operational experience, the Merlin engine is on track to become one of the most frequently flown of the commercial space age.

The Merlin may also be on track to achieve one of Musk's goals, namely transforming the commercial space arena and opening up space to the general public. The fact the Merlin has successfully flown in space and is in production (and will, by all accounts, be so for many years to come) means the commercial space sector has a common off-the-shelf, flight-ready engine readily available that could be key to the development effort of a first-generation reusable orbital system. For suborbital start-ups with a suborbital vehicle on the runway/launch pad, the financial and technical challenges of designing and testing an engine to propel it remain daunting in the extreme. But, being able to purchase an engine, such as the Merlin, with a known performance range and a strong flight history could lower the bar from nearly impossible to just really hard. So, rather than struggling through a potentially interminable and expensive engine-development program, engineers could instead focus on all the other myriad challenging issues such as airframe, re-entry, life-support systems, emergency-egress systems, and landing.

Selling Merlins also makes sense from the perspective of SpaceX's low-cost access to new markets. For example, what if SpaceX sticks to launching Falcon rockets and doesn't pursue the development of other platforms? In this case, the company might reckon that it's better to have its engines powering alternate space launch systems through partnership agreements than risk being left out in the cold in the event that another company produces a vehicle with their own engines. Unlikely, but it could happen. But, by offering the Merlin, a product that SpaceX has already developed, SpaceX could position itself to become an integral player in many key areas of the space tourism industry, and it can do this without much additional risk or expense.

The spirit of collaboration has been echoed by others in the commercial spaceflight business. For example, at a ceremony unveiling the design for WhiteKnightTwo in 2008, Virgin Galactic President, Will Whitehorn, said his company intended to make the SpaceShipTwo technology "open source architecture like Linux" and that Virgin Galactic would welcome approaches from outside sources in developing new applications. Perhaps a similar statement from Musk on the subject of the eventual availability of the Merlin engine might open the door to more interesting possibilities in the future.

Thanks to the success of the June 2012 Merlin 1D test, SpaceX has removed another obstacle on the road to a more reliable and flexible launch architecture and has moved one step closer to achieving Musk's goal of reducing the cost of delivering a pound of payload to LEO to US$1,000 or less. More steps will be taken as SpaceX uses its Merlins to launch more payloads as the company begins to fulfill its contract of a dozen ISS cargo delivery missions between 2012 and 2015.

Given the number of cargo missions slated to fly and given the potential for the engine playing a role outside the Falcon family of launch vehicles, it is very likely the Merlin may be destined for an exceptionally long lifespan. After all, the aforementioned NK-33 engines have waited more than 40 years to fulfill their potential, with perhaps another decade or more to go until they have all flown. In terms of longevity, the all-time winner is undoubtedly the venerable RL-10, which was ground-tested in 1959, first flown in 1963, powered the DC-X test vehicle, remains the heart of the Centaur upper stage, and appears to have a credible future as a deep-space engine. It seems we are safe in taking the long view of things. After all, the Chevrolet small block found ultimate greatness not just through the genius of its creators or in its original design specifications, but because the design itself was capable of evolving and being adapted by those who were not even born when the original engine first roared to life.

4

Rise of the Falcon

"You came in that thing? You're braver than I thought."
Leia Organa upon seeing the *Millennium Falcon* for the first time

In the sci-fi classic, *Star Wars*, the iconic *Millennium Falcon* spacecraft (Figure 4.1) is commanded by Harrison Ford's character Han Solo. Elon Musk must have been a fan, because SpaceX's fleet of Falcon rockets is named after George Lucas's fictional spaceship, although the two vehicles bear little resemblance and their roles have little

4.1 *Millennium Falcon.* Courtesy: *http://img690.imageshack.us/img690/7137/falconw.jpg*

in common; whereas the *Millennium Falcon* was designed to go tearing around the galaxy looking for trouble, its namesake was designed to provide breakthrough advances in reliability, cost, and time to launch.

FALCON 1

With funds stretched tight after founding SpaceX, Musk decided to develop a small practical space launcher. That launcher turned out to be 21.3 meters tall, 1.7 meters in diameter, featuring a rocket-grade kerosene (RP-1)-fueled two-stage rocket capable of boosting about 0.6 tonnes to low Earth orbit (LEO). Its first stage is boosted by the in-house-developed Merlin engine, while the second stage is powered by the Kestrel engine. The first stage, which is helium-pressurized and designed to be recovered at sea, is a pressure-assisted stabilized graduated monocoque aluminum design that uses a common bulkhead between its aft kerosene tank and its forward liquid-oxygen (LOX) tank. The expendable second stage, which is also helium-pressurized,[1] is fabricated from aluminum; the plan had been to use lighter aluminum–lithium, but SpaceX was unable to secure the metal.

Falcon 1's mission architecture is as simple as it is elegant. A single SpaceX Merlin engine powers Falcon 1's first stage and, after engine start, the vehicle is held down until all systems are verified to be functioning normally, at which point the vehicle blasts off (Figure 4.2). Stage separation is achieved via dual initiated separation bolts and a pneumatic pusher system and the retrievable first stage returns by parachute to a water landing, where it is picked up by ship in a procedure similar to that employed to recover the Shuttle's solid-rocket boosters (SRBs). Incidentally, SpaceX's parachute recovery system is built by Airborne Systems Corporation, which also built the Shuttle booster recovery system.

Selling Features

For those buying a Falcon flight, a Collision Avoidance Maneuver (CAM) is provided as a standard service – no extra charge. The CAM – if it's required – can be performed using the Reaction Control System (RCS) thrusters which are tilted forward 20° and positioned to minimize gas impingement on the spacecraft while still providing adequate separation. Another service provided to customers is a restart capability, which provides the flexibility required for payload insertion into orbit with varying eccentricities and the deployment of multiple payloads into different orbits. The company advertises the following orbital insertion accuracies: inclination $\pm 0.1°$, perigee ± 5 km, and apogee ± 15 km.

Falcon 1's guidance, navigation, and control (GNC) system includes a ruggedized

---

[1]   Helium tank pressurization is achieved by composite over-wrapped inconel tanks manufactured by the Arde Corporation – the same model as used in Boeing's Delta IV rocket.

4.2 Static test firing of the Merlin 1C engine on the first stage of the Falcon 1 Flight 4 vehicle, Omelek Island in the Kwajalein Atoll, on September 20th, 2008. Courtesy: SpaceX

flight computer and an inertial measurement unit (IMU) backed by a global positioning system (GPS) receiver which is flown for navigation updates. The GNC system also includes an S-band telemetry system, an S-band video downlink, and a C-band transponder.

SpaceX provides a standard payload separation system for Falcon 1 but the company can also integrate a separation system chosen and provided by the payload provider. Payload separation, which is initiated non-explosively by separation springs that impart separation velocity, is a timed event referenced to the second-stage burnout. Other options provided to the customer include spinning up the payload, which can be spun up to six revolutions per minute (RPM) at separation. Attitude is another option that can be customized, with attitude and rate accuracies that include plus or minus two degrees of roll, half a degree of pitch and yaw, and one-tenth of a degree of body rate.

Another selling feature of Falcon 1 is its reliability. SpaceX took heed of the lessons learned from previous launch vehicle failures (most of which were attributed to engines, avionics, or stage-separation failures) and built a robust propulsion system with a redundant ignition system and matched this with a vehicle that features a state-of-the-art avionics system. In fact, one of the primary goals when designing Falcon 1 was to ensure reliability was not compromised – an approach that required some key choices. Among these was designing the first stage to be recovered and reused, which meant this part of the vehicle had to have significantly higher margins than an expendable stage; during the testing of the first stage, SpaceX subjected a first stage to more than 190 cryogenic pressure cycles with no evidence of fatigue. Another decision that featured in the reliability matrix was to minimize risk during the propulsion and separation phase – a goal that was achieved by designing Falcon 1 with the minimum number of engines in serial. Off-nominal propulsion events are further reduced by SpaceX s launch operation procedures which involve holding down the first stage after ignition but, prior to release, to watch engine trends; if an off nominal condition is detected, an autonomous abort is conducted. More reliability is achieved by reducing the number of failure modes by minimizing the number of separate subsystems. For example, Falcon 1's first-stage thrust vector control (TVC) system makes use of the pressurized fuel, RP 1, through a line tapped off of the high-pressure RP side of the pump to power the TVC – a design that not only eliminates the separate hydraulic system, but also eliminates the failure mode associated with running out of pressurized fluid.

All of the Falcon 1's design decisions have been subject to exhaustive testing, ranging from component-level qualification and workmanship testing, structures load and proof testing, to flight system and propulsion subsystem-level testing. Adding another layer of ruggedization to their vehicle, SpaceX has tested beyond the margins of environmental extremes and has conducted stage and fairing separation tests for off nominal events such as geometrical misalignment, anomalous pyrotechnic timing, and sequencing.

The final layer of reliability is a robust launch operation. In common with government space agencies, SpaceX's countdown sequence is fully automated, with thousands of checks made prior to vehicle release.

LAUNCHING A PAYLOAD

For customers who have a satellite they would like to launch, the first step is to fill out a Payload Questionnaire. The Payload Questionnaire allows SpaceX to assess mission feasibility, define mission requirements, and fine-tune launch countdown procedures. The questionnaire requires the customer to provide detailed information about their payload, such as a mathematical model, which is needed before mission integration can be performed. In addition to the mathematical model, the customer must also provide an interface control document that describes all mission-specific requirements. An environmental statement is also required, as is a specification of the radio frequencies that will be transmitted by the payload during ground processing and launch operations. This latter requirement goes into some detail, requiring the customer to list not only the individual frequencies, but also the names and qualifications of the personnel who will operate the radio frequency systems, the duration of transmission, and the frequency bandwidths. Details of the payload design, including graphics and configuration drawings showing dimensions, are also required together with the procedures that are planned at the launch site operations; this information is then passed on to government agencies and range safety.

An important reference document that helps customers complete the Payload Questionnaire is the *Falcon Users Guide*, a publication not that dissimilar to a car user's manual, which provides a detailed set of instructions of what can and cannot be accommodated. In addition to the standard performance characteristics of Falcon 1, the manual also explains very specific information about payload integration, which informs potential customers about the options available to them when launching a payload. For many customers, it's important to understand the operating conditions that will affect their payload not only during launch, but also while the payload is in orbit. For example, a customer hoping to launch a payload that contains consumables will require a different set of operating conditions to one with no consumables.

Once a payload has been accepted for launch on board Falcon 1, the process of installing the payload follows a general sequence of events. First, each payload must interface with the launch vehicle by means of a Payload Attach Fitting (PAF), which can be modified to accommodate customer needs. Next, before the vehicle is actually shipped to the launch site, a mechanical fit check that includes electrical connector locations has to be configured with the spacecraft. This check is conducted by SpaceX personnel during the payload integration process. The payload also has to be checked for electrical design interface for ground and flight operations. SpaceX prefers that satellites be powered off during launch, although satellites can be launched while powered on as long as they aren't transmitting, due to the potential for interference.

The customer must also consider the environment their payload will be exposed to from when it leaves the customer's location until it is released in orbit. Much of this information is contained in SpaceX's Interface Control Document (ICD), which defines the various environments. For example, the first environment to consider is the transportation environment, which is encountered by the payload during its journey from the payload-processing hangar to the launch pad, which may be accomplished by wheeled vehicle and ocean vessel. Since SpaceX doesn't control ambient temperature, humidity, and cleanliness during this journey, it's up to the customer to ensure that their payload transportation containers are designed to protect the payload until it is finally removed from the container in the environmentally controlled payload-processing facility.

The next stage is looking after the payload while it is in the clean room before being stowed away within the fairing. The clean room is an air-conditioned facility maintained at $21 \pm 5.5°C$ and a humidity level between 30% and 60%. The payload remains in an air-conditioned environment after being encapsulated, thanks to air being provided via a flexible duct system to a fairing port configured to direct air into the fairing. From this stage, except for a short break to move the payload from the clean room to the vehicle, the payload remains in an air-conditioned environment until launch.

Launch is the toughest environment the payload is subjected to during its journey from customer to orbit. In fact, the launch loads to which the payload is subjected in the first 10 minutes of the payload's climb to orbit drive much of its structural design and mass. SpaceX is acutely aware of the stresses placed on a payload during launch and has gone to great lengths to reduce Falcon 1's launch loads. However, despite the decision to forego the use of solid-rocket boosters (a culprit in imposing high launch loads), there is no escaping the shock loads that occur during every flight. First there is the shock of the hold-down release of Falcon 1 at lift-off followed shortly after by stage separation. While these shocks are fairly negligible, the shock of fairing separation can be punishing due to the distance and number of joints over which the shocks travel and dissipate.

Less punishing is the radiofrequency (RF) environment, but customers are still required to ensure that components sensitive to an RF environment are compatible with the launch pad environment, which is characterized by frequencies governing command and destruct, tracking transponder, vehicle launch telemetry, GPS, UHF, C-band, and S-band.

Other considerations include spacecraft fueling, which SpaceX accommodates as a non-standard service, and electrical power supply, which requires the customer to provide the necessary cables to interface the payload with the payload-processing room. There is also the issue of monitoring the payload once in orbit. During test and launch operations, SpaceX provides one console for the customer in the SpaceX command center and stations for up to five other payload support personnel during launch operations, either in the payload-processing area or in other facilities. Once SpaceX is happy with all the information and with the payload specification provided by the customer and once the customer (the payload provider in SpaceX parlance) is happy with the services provided by SpaceX, the customer is assigned a Mission Manager, who serves as a point of contact from contract award through launch. It's the Mission Manager who assesses the launch vehicle capabilities against payload requirements, conducts mission design reviews, and attends to the myriad teleconferences and integration meetings (Table 4.1) which are required before launching a payload into orbit. It is also the Mission Manager's job to coordinate other administrative aspects such as range safety integration and mission-required licensing. Once the payload arrives at the launch site, the physical accommodation for the spacecraft is turned over to the Payload Integration Manager, although the Mission Manager continues to manage the customer interface at the launch site.

Once the customer's payload has made it to the launch site, SpaceX makes pre-launch operations as simple and as streamlined as possible – a sequence of events that begins 18 days prior to launch. Before a Flight Readiness Review (FRR) can be completed, payload attachment and fairing encapsulation must be completed. The process normally takes less than 24 hours and begins by integrating the payload on the adapter in the vertical configuration, followed closely by fairing encapsulation. Once fully encapsulated, the system is rotated horizontally and integrated to the second stage. Once this has been completed, post-mate checkouts can be conducted, which in turn are followed by a FRR. Once the FRR is completed, Falcon 1 is rolled out to the pad. Six days prior to launch, the integrated payload and launch vehicle are positioned vertically using the Falcon 1 Launch Vehicle transporter. Final system close-out, fueling, and testing are then completed. Twenty-four hours prior to

Table 4.1. Generic launch integration process.

Timeline	Activity
T–8 months	‣ Estimated payload mass, volume, mission, operations and interface requirements ‣ Safety information; design information such as battery, ordnance, propellants, and operations ‣ Mission analysis summary provided to customer within 30 days of contract·
T–6 months	‣ Final payload design, including mass, volume/structural characteristics, mission, operations, and interface requirements ‣ Payload to provide test-verified structural dynamic model
T–4 months	‣ Payload readiness review for Range Safety – includes launch site operations plan and hazard analyses
T–3 months	‣ Review of payload test data verifying compatibility with Falcon 1 environments ‣ Coupled payload and Falcon 1 loads analysis completed ‣ Confirm payload interfaces as built are compatible with Falcon 1 ‣ Mission safety approval
T–4 to 6 weeks	*System Readiness Review (SRR)* ‣ Pre-shipment reviews ‣ Launch site verified; range, regulatory agencies, launch vehicle, payload, people, and paper in place and ready to begin launch campaign
T–2 weeks	Payload arrival at launch location
T–8 to 9 days	Payload mating to launch vehicle and fairing encapsulation
T–7 days	*Flight Readiness Review (FRR)* ‣ Review of launch vehicle and payload checkouts in hangar ‣ Confirmation of readiness to proceed with vehicle rollout
T–1 day	*Launch Readiness Review (LRR)*
Launch	
T+4 hours	Post-Launch Reports – Quick Look
T+4 weeks	Post-Launch Report – Final Report

launch, the Launch Readiness Review (LRR) is held. Once the launch approval is given, the 24-hour countdown begins.

Following launch (Figure 4.3), the vehicle follows a flight profile that varies depending on the trajectory. For example, for direct injected missions, first-stage burnout/stage separation occurs at T+169 seconds at an altitude of 297,000 feet. This event is followed by second-stage ignition at T+174 seconds at an altitude of 324,000 feet. Fairing separation occurs 20 seconds later at an altitude of 429,000 feet and second-stage burnout occurs at 552 seconds at an altitude of 1,333,200 feet with payload deployment occurring at T+570 seconds.

4.3 Falcon 1 launch. Courtesy: SpaceX

FALCON 1 DEVELOPMENT

In keeping with the SpaceX mode of operation, development of Falcon 1 was rapid, the fabrication of a proto-vehicle beginning in early 2003. Less than a year later, on December 3rd, 2003, after driving cross-country on its custom-built transport trailer, SpaceX unveiled the proto-vehicle in Washington, DC, parking it on the street in front of the Federal Aviation Administration (FAA) building. Musk used the occasion to pronounce that his company planned a sequel to Falcon 1 by building a more powerful 3.7-meter-diameter Falcon 5 to be powered by five Merlin engines. Initial pricing for Falcon 1 was set at US$6 million, while Falcon 5, designed to haul 4.5 tonnes to LEO, was listed at US$12 million.

In September 2004, SpaceX won a Defense Advanced Research Projects Agency (DARPA) contract that included a Falcon 1 launch from Omelek Island in Kwajalein Atoll in the Marshall Islands. This was a month before the company erected its first Falcon 1 at the company's SLC 3W launch pad at Vandenberg Air Force Base on October 5th, 2004. While SpaceX built a backlog of missions, including one to launch TacSat-1, a US Navy microsatellite, and another to orbit a test payload for Bigelow Aerospace, the company had to wait for final development of the Merlin engine. They also had to complete Falcon 1 development, which was finally achieved with a series of structural tests on March 31st, 2005.

With these milestones reached, SpaceX was ready to launch TacSat-1, but military bureaucracy got in the way because the Air Force didn't want SpaceX to launch until a Titan 4 flew from nearby SLC 4E. After repeated delays pushed the Titan launch back, Musk decided to fly the first Falcon 1 from Kwajalein instead. So, in June 2005, SpaceX ferried the Falcon launch equipment to the Marshall Islands, followed by the first Falcon 1 vehicle a month later.

The first launch attempt at Omelek on November 25th, 2005, didn't go well, the launch effort being scrubbed after a ground-supply LOX vent valve allowed the LOX supply to boil off. A second attempt was made on December 19th, 2005, but this was delayed by high winds. Worse was to come when the first-stage fuel tank buckled during fuel draining when the fuel pressurization system suffered a controller failure. The damaged first stage had to be shipped to Los Angeles for repair and was replaced with the second Falcon 1's first stage.

Less than two months later, SpaceX was ready to try again. On February 9th, 2006, the company completed a hot-fire test at the Omelek pad with the new first stage. Unfortunately, a second-stage propellant leak was discovered during the testing process, which nixed the February launch attempt. The company shipped the troublesome second stage back to Los Angeles and replaced it with the second Falcon 1's second stage. On March 18th and 23rd, 2006, SpaceX ramped up for a fourth launch attempt by performing hot-fire tests using the reconfigured vehicle.

Having successfully performed the hot-fire tests, everything seemed to be on track for a successful inaugural launch on March 24th, 2006. After a 22:30 GMT lift-off, the Falcon rose from its pad and ascended in what appeared to be a clean, stable ascent. The launch proceeded according to the script until about 25 seconds into flight, when a fire just above the engine cut into the first-stage helium pneumatic system, causing an engine shutdown at T+34 seconds. Following the shutdown, Falcon 1 rolled and fell into the ocean.

Since Falcon 1 is equipped with an engine cut-off range safety system rather than destruct charges, the vehicle fell more or less intact on a reef not far from the launch site (Falcon 1's payload, an experimental microsat built by US Air Force Academy students, crashed through the roof of a shop building on the island). SpaceX immediately went to work troubleshooting the problem. A week later, in a NPR interview, company vice president, Gwynne Shotwell, informed the news media that the leak had been caused by a procedural error rather than a hardware failure of Falcon 1. Four months after the failed launch, SpaceX reported the findings of a DARPA "Falcon Return to Flight Board", which revealed that a kerosene fuel leak had begun 400 seconds before lift-off, when the propellant pre-valves were opened. When the Merlin main engine had started at lift-off, the leaking fuel had ignited.

On August 18th, 2006, SpaceX announced it had been selected by NASA to demonstrate delivery and return of cargo to the International Space Station (ISS) as part of the agency's Commercial Orbital Transportation Services (COTS) competition (see Chapter 2). In addition to SpaceX, NASA signed agreements with Rocketplane Kistler (RpK) and Alliant Techsystems (ATK), but later terminated the agreement with RpK due to insufficient private funding. As a private spaceflight vendor in the COTS competition, SpaceX was required to compete in four service areas:

- Capability level A: External unpressurized cargo delivery and disposal
- Capability level B: Internal pressurized cargo delivery and disposal
- Capability level C: Internal pressurized cargo delivery, return, and recovery
- Capability level D: Crew transportation

The COTS program was the product of some imaginative and innovative thinking by NASA. In essence, the agency reckoned that companies in a free market had a better chance of developing and operating a crew and cargo system more efficiently and affordably than a government bureaucracy. Another key factor motivating NASA's decision to turn over spacecraft development to the private sector was the impending retirement of the Space Shuttle. With the Shuttle due to retire by the end of 2010 and NASA's own Crew Exploration Vehicle (CEV) still in the design phase, the agency was relying on a commercial cargo service being developed to avoid having to purchase orbital transportation services on foreign spacecraft such as the Russian Federal Space Agency's Soyuz and Progress spacecraft or the European Space Agency's (ESA) Automated Transfer Vehicle (ATV). As part of Musk's agreement with NASA, SpaceX would execute three flights of its Falcon 9 rocket carrying the Dragon spaceship. The missions were scheduled to occur in the late-2008-to-2009 time period and would culminate in demonstrating delivery of cargo to the ISS and safe return of cargo to Earth.

As SpaceX went to work preparing for its next Falcon 1 launch attempt, it also announced revised design information for its launch vehicle, providing details on its website of a new Merlin 1C-powered Falcon 1e rocket that would be 5.53 meters taller and 11.36 tonnes heavier than the original Falcon rocket. Falcon 1e, which was expected to enter service after 2009, would be able to haul 25–30% more payload than Falcon 1.

The next attempt to launch Falcon 1 was scheduled to take place early in 2007. After being erected at Omelek in mid-January 2007, a late-January hot-fire test was postponed when the vehicle's second-stage engine failed a slew test during the countdown with the result that the vehicle was moved back into its hanger. On March 15th, 2007, SpaceX performed a successful static test ignition of the Falcon 1 first-stage Merlin engine. This was followed by a scrubbed launch attempt on March 19th, 2007. Then, two days later, Musk's Falcon 1 failed to reach orbit after flight control was lost after just two minutes into the vehicle's second-stage burn.

While not completely successful,[2] the flight achieved a number of milestones,

---

[2]   At launch, Falcon 1 weighed 27.526 tonnes. First-stage burnout occurred 168 seconds after lift-off at an altitude of 75 kilometers and a velocity of 2.6 kilometers per second. The second-stage Kestrel engine ignited five seconds after first-stage cut-off, beginning a planned burnout of about 415 seconds intended to insert the stage into an initial 330 × 685 kilometer orbit about 585 seconds after lift-off. Ultimately, the second stage achieved a suborbital velocity of about 5.1 kilometers per second, reaching a maximum altitude of 289 kilometers.

including passing through Max Q, completing a full first-stage burn, stage separation, second-stage ignition, and jettisoning of the payload fairing. The launch also demonstrated SpaceX's operational responsiveness thanks to a launch abort that stopped the main engine start sequence. The abort, which was caused by a low chamber pressure reading resulting from lower-than-planned kerosene fuel temperatures, required SpaceX crews to drain and reload some of the first-stage fuel. The glitch allowed the SpaceX crew to demonstrate to DARPA officials just how quickly they could reload propellant, the whole exercise taking less than an hour before restarting the count.

The reason for the partial failure was captured on the on-board video broadcast, which showed the second-stage engine bell brushing against the side of the inter-stage at stage separation. The video also revealed an oscillatory motion developing during the last minute of controlled flight, just before roll control and telemetry were lost. Six days later, Musk explained that LOX sloshing had caused the oscillation; the LOX slosh frequency had coupled with the TVC system in a way that amplified the oscillation until flight control was lost.

Bankrolling the partially successful Falcon 1 flight was the DARPA under the auspices of the DARPA/USAF Falcon program. The primary aim of the US$7 million mission had been to gather flight data for the US Department of Defense, a major SpaceX customer. On board Falcon 1 was a 50-kilogram experimental satellite that had been destined for a circular 685-kilometer orbit.

In April 2008, SpaceX revealed details of a beefed-up Merlin 1C, capable of producing more than 56 tonnes of sea-level thrust – a capability that translated into a one-tonne LEO payload for Falcon 1e (by comparison, Falcon 1's LEO payload was now advertised as 0.42 tonnes). Later that year, on August 3rd, the third Falcon 1 rocket failed shortly after lifting off from Omelek Island, the cause of the failure being attributed to residual thrust produced by the Merlin 1C first-stage engine that caused the stage to re-contact the second stage immediately after stage separation. Lost with Falcon 1 were the USAF's Trailblazer satellite, NASA's Nanosail-D solar sail experiment, and NASA's PreSat experiment. Echoing the second attempt, the third launch effort also suffered a countdown abort, this one 34 minutes before launch. SpaceX crews, demonstrating their efficiency once again, recycled the count in 23 minutes.

FALCON 1'S FOURTH AND FIFTH FLIGHTS

Less than two months later, on September 28th, SpaceX was ready to go again. This time, the launch went as advertised and the company's fourth Falcon 1, carrying a 165-kilogram payload mass simulator, reached a 330 × 650 kilometer × 9° orbit. There were a few minor deviations from the plan, the first being a slightly lower orbit than the 330 × 685 kilometers that had been the target. Another minor departure from the flight profile was a slightly early shutdown of the second stage, but this was compensated for by the Kestrel second-stage engine performing a test of its restart capability in space. But these were minor issues; the major objective had been achieved and the Falcon 1 booster had redeemed itself with an electrifying launch that put an exclamation point on six years of hard work and disappointment for SpaceX. The mission logo for the launch, known as Flight 4 in SpaceX parlance, included two four-leaf clovers symbolizing the end of the rocket's string of bad luck.

After flights plagued by delays and last-second aborts, the successful launch of Falcon 1 marked a major milestone and accomplished a feat that only a handful of countries had achieved. Priced at just US$7.9 million per flight, Falcon 1 was three times less expensive than its US competitors in the commercial launch market and its Russian, Chinese, and Indian rival launch vehicles. After the inevitable setbacks and delays that are a fact of life when trying to develop a new rocket, SpaceX had demonstrated that it had the patience and the knowledge to adapt and innovate after every launch failure and, in so doing, deliver on its promise of putting the company on the map of commercial spaceflight. Having demonstrated they could get into orbit, the company's next objective was to show they could repeat the feat.

On July 14th, 2009, SpaceX launched its fifth Falcon 1, boosting the RazakSAT, a Malaysian-government Earth-observation imaging satellite, into orbit. Following a 2-minute 40-second burn, the first stage fell away and the second-stage Kestrel engine ignited and completed its first burn about 9 minutes 40 seconds after lift-off, boosting the stage and 180-kilogram payload toward a 330 × 685 kilometer parking orbit. The launch marked the final original Falcon 1 on the SpaceX manifest and the first launch of a live satellite.

The smooth RazakSAT flight underscored SpaceX's position in the tough world of commercial spaceflight. In just seven years of existence, it had positioned itself as a force to be reckoned with – an achievement that was only possible because it was a private company. Because it hadn't been encumbered by government red tape, Musk had been able to develop his company as he wanted – as long as the money was there, which it was thanks in part to NASA and its COTS program.

THE END OF FALCON 1

After so much work and money developing and flight-proving Falcon 1, the expectation was that SpaceX would begin to exploit its new vehicle's operational status and aggressively market and sell its new launch service. But, since the success of RazakSAT, no Falcon 1 launch has occurred. Go to the SpaceX website and you can still find descriptions of Falcon 1, but if you go to the bottom of the page, you will read that SpaceX no longer sells the rocket. Instead, small, one-tonne-class payloads will be launched on piggyback rides on the Falcon 9 launcher.

For SpaceX to spend six years and millions of dollars to develop a launch system only to abandon it just as success and profit are at hand is difficult to understand. Was the development of Falcon 1 as a commercial launch system never intended in the first place? Perhaps the justification of the development of the launcher was simply to flight-qualify components such as the Merlin 1 engine for eventual use on the Falcon 9 launcher? What *is* known is that customers who require a low-cost option for launching their small payloads need to look elsewhere. But where? One of the companies that *does* launch small satellites is the Orbital Sciences Corporation ("Orbital"), whose Taurus launch vehicle costs around US$50 million to US$70 million, but their recent record of unreliability (e.g. the Glory satellite launch failure) makes it a risky proposition.

Perhaps equally significant is the fact that, after investing in the development of a new, unproven company that was offering a low-cost launch vehicle, SpaceX's defense customers, who were banking on a quick, inexpensive capability to launch small satellites, saw their support of Falcon 1 evaporate. Why? Well, it would seem that SpaceX, having succeeded in developing Falcon 1, dropped the vehicle and instead focused on the promotion of more ambitious goals such as developing the Falcon Heavy and manned missions to Mars.

In an interview with NASASpaceflight.com, when the company was asked whether there was still a future role for Falcon 1/1e, SpaceX's communications director Kirstin Brost Grantham said the company's plans were for small payloads to be served by flights on Falcon 9, utilizing excess capacity. In other words, SpaceX doesn't see a future for Falcon 1e, which is logical given that Falcon 9 can launch small satellites just fine as secondary payloads. So, Falcon 1 has probably come to an end. It's legacy? Well, it depends on your viewpoint. From the perspective of launch successes, Falcon 1 had one of the all-time worst orbital launch vehicle records in history; it failed three times in five attempts and managed to send only one satellite into orbit and even that satellite failed to function properly. It's a terrible launch record and one that is only rivaled by India's GSLV[3] and Iran's Safir (which succeeded three times out of four) among modern rockets that have flown more than twice. From a legacy perspective, Falcon 1 represents solid engineering, the results of which were transferred to the development of Falcon 9, which, to date, is three for three. Given Falcon 9's success rate and SpaceX's practice of "build a little, fly a little", there is no reason to think that future Falcon flights shouldn't continue this success rate.

---

[3]   The Geosynchronous Satellite Launch Vehicle (GSLV) is an expendable launch system operated by the Indian Space Research Organization (ISRO) developed to launch satellites into geostationary orbit . Since its first launch in 2001, there have been seven GSLV launches: two successful, one partially successful, and four failures. Flight number eight is scheduled for early 2013.

5

Falcon 9 and Falcon Heavy: Life after the Space Shuttle

"The space shuttle has changed the way we view the world and it's changed the way we view our universe. There's a lot of emotion today, but one thing's indisputable. America's not going to stop exploring."

STS-135 Space Shuttle Commander Chris Ferguson

It was a hot July day on Florida's Space Coast as nearly a million spectators gathered along the beaches and causeways to watch history in the making. Just as launch looked imminent, a last-minute glitch held the clock at T–31 seconds, but the issue was quickly resolved and the clock began counting down the final seconds. With less than a minute remaining in the launch window, *Atlantis*'s three main engines roared to life and the twin solid-rocket boosters thundered. For the 33rd and final time, *Atlantis* rose (Figure 5.1) majestically from the launch pad on a plume of fire and parted the clouds on its way to the International Space Station (ISS) and to its place in history. The 11:29 EDT lift-off on July 8th, 2011, marked the last time a Space Shuttle would climb from the Kennedy launch complex.

Less than 13 days later, on July 21st, for one last time, *Atlantis* made a long, steep turn, lined up with the runway, and landed in the half-light before dawn at NASA's Kennedy Space Center (KSC) in Florida. Wheel stop came at 05:58 EDT, after a flight of 12 days, 18 hours, 28 minutes, and 55 seconds. And that was that. After 135 flights in 30 years, and having traveled more than 800 million kilometers in orbit, the Space Shuttle era was history. *Atlantis* alone made 33 flights, carried 191 astronauts, spent 307 days in orbit, circled Earth 4,848 times, and put more than 200 million kilometers on the clock. With *Atlantis* on the ground, 2,300 Shuttle workers received layoff notices and more than 8,000 people who worked for NASA or its contractors in the Shuttle program lost their jobs. It was a quiet ending to a program that was supposed to have made spaceflight affordable, safe, and routine. Instead, it proved risky and expensive, with the average cost of a Shuttle flight estimated at half a billion dollars.

The final landing of *Atlantis* left NASA with no spacecraft to replace its capability to send its astronauts to orbit. As a stopgap measure, NASA signed a US$763 million contract for 12 Russian rocket rides (that's more than US$63 million per

5.1 Space Shuttle *Atlantis*. Courtesy: NASA

flight!) from 2014 through 2016. By that time, the space agency hopes at least one of the four private companies it's seeding with cash will demonstrate a crew-ready launcher. Leading the way is SpaceX with its Falcon 9.

FALCON 9

Falcon 9 (Figure 5.2) is a two-stage rocket powered by liquid oxygen (LOX) and rocket-grade kerosene (RP-1). In common with Falcon 1, it was designed from the ground up by SpaceX for cost-efficient transport of satellites to low Earth orbit (LEO) and geosynchronous transfer orbit (GTO), and for sending the Dragon spacecraft, carrying cargo and/or astronauts, to orbiting destinations such as the ISS.

It's a "Made in America" rocket with all structures, engines, avionics, *and* ground systems designed, manufactured, *and* tested in the US by SpaceX. Designed to one day carry crew, Falcon 9, with a Dragon spacecraft perched on top, is 48.1 meters tall. Developed from a blank sheet to first launch in just four and a half years (November 2005 to June 2010), Falcon 9 is capable of producing 1,000,000 pounds of thrust in a vacuum. The cost? Less than US$300 million. The vehicle features cutting-edge technology and a simple two-stage design to limit separation events; with nine engines (hence the number "9" in the name) on the first stage, Falcon 9 can still safely complete its mission in the event of an engine failure.

5.2 Falcon 9 launch. Courtesy: SpaceX

First and Second Stages

The tank walls (Figure 5.3) are made from an aluminum lithium alloy that SpaceX manufactures using friction-stir welding, which is the strongest and most reliable welding technique available. Powering the first stage is a cluster (Figure 5.4) of nine SpaceX Merlin regeneratively cooled engines. Connecting the lower and upper stages is the inter-stage, a composite structure with an aluminum honeycomb core and carbon-fiber face sheets. The separation system is pneumatic – a system proven on its predecessor, Falcon 1.

5.3 Rocket tank production. Courtesy: SpaceX

The Falcon 9 second-stage tank is a shorter version of the first-stage tank and uses many of the same tooling, material, and manufacturing techniques – a measure that results in significant cost savings in vehicle production. Powering the upper stage is a single Merlin engine, capable of restart thanks to dual redundant pyrophoric igniters using triethylaluminum–triethylborane (TEA–TEB). The rocket is a reliable system, in part because it only has only two stages, which limits problems associated with separation events. This reliability is enhanced by the incredibly advanced avionics systems, which featured a hold-before-release system. It's a capability required by commercial airplanes, but not implemented on many launch vehicles. Here's how it works: after the first-stage engine ignites, Falcon 9 is held down and not released for flight until all propulsion and vehicle systems are confirmed to be operating normally; if any issues are detected, an automatic safe shutdown occurs and propellant is unloaded.

5.4 Engine cluster. Courtesy: SpaceX

FLIGHT #1

Capable of lifting payloads of 10,450 kilograms to LEO and 4,450 kilograms to GTO, Falcon 9 launched – after several delays – on its maiden flight from Cape Canaveral Air Force Station on June 4th, 2010, at 14:45 EDT. Lift-off came 3 hours and 45 minutes into a four-hour launch window because of tests conducted on the rocket's self-destruct system and a last-second abort caused by a higher-than-expected pressure reading in one of the engines. SpaceX engineers resolved the issue and recycled the countdown to the T–15-minute mark after concluding the engine was in good shape.

Falcon 9, Flight #1 Launch Timeline

T+00:00:06 Lift-off
T+00:01:13 Falcon 9 approached maximum dynamic pressure (Max Q)
T+00:01:56 Plume behind the vehicle expanded as the atmosphere thinned
T+00:03:06 Stages separated
T+00:03:34 Second-stage ignition
T+00:08:50 Second-stage engine shutdown
T+00:09:04 Falcon 9 achieved Earth orbit, delivering a dummy payload, a structural test article representing the company's planned Dragon space station cargo module

Telemetry from Falcon 9 confirmed the vehicle had achieved an orbit of almost exactly 250 kilometers. If you consider that two-thirds of rockets introduced in the past 20 years have had an unsuccessful first flight, the maiden Falcon 9 flight couldn't be considered as anything other than 100% successful, but Musk and his team of engineers weren't the only ones breathing a sigh of relief as Falcon 9 sailed above their heads; NASA, which had entered into contracts with SpaceX totaling US$1.6 billion for deliveries to the ISS, no doubt also started feeling a lot more comfortable.

The successful Falcon 9 launch not only boded well for the US shift in national space priorities, turning launches to LEO to the private sector while NASA focuses on deep-space exploration, but also vindicated Musk's approach by demonstrating that a small, new company could make a difference. Talking to the media following the launch, Musk acknowledged NASA's help, but also pointed out that the first Falcon 9 heralded the dawn of a new era of spaceflight which will be increasingly marked by combined commercial and government endeavors, with commercial companies playing an increasingly significant role. But, while the flight was no doubt historic, it was difficult to judge whether it was a game-changer. After all, SpaceX couldn't have succeeded without significant government money and the promise of a healthy supply contract and they had plenty of technical support from NASA. Having said that, the first Falcon 9 flight represented a step away from the duplicative and wasteful government's way of doing business and no doubt provided a stimulus for other companies to follow in SpaceX's path. Of course, other companies had also followed a similar route to SpaceX and failed. For example, the Orbital Sciences Corporation ("Orbital") was the latest and greatest spaceflight company in the early 1990s, pronouncing they were going to reduce the cost of going into space but, eventually, they were absorbed into the system and became just another highly priced aerospace company. On the subject of cost, it was really a case of comparing vehicles in terms of dollars per kilogram lifted. Compared to the Space Shuttle, SpaceX had a good case for saying they could launch for a tenth of the cost but, compared to the Atlas V551 or the Delta IV Heavy, that ratio was probably closer to a fifth. Unlike the Delta and Atlas, however, which only launch once or twice per year, SpaceX's mass production and paralleling of systems mean that economies of scale creep in, allowing Musk to advertise a Falcon 9 launch for just US$54 million.

FLIGHT #2

"When Dragon returns, whether on this mission or a future one, it will herald the dawn of an incredibly exciting new era in space travel. This will be the first new American human capable spacecraft to travel to orbit and back since the Space Shuttle took flight three decades ago. The success of the NASA Commercial Orbital Transportation Services (COTS)/Commercial Resupply Services (CRS) program shows that it is possible to return to the fast pace of progress that took place during the Apollo era, but using only a tiny fraction of the resources. If COTS/CRS continues to achieve the milestones that many considered impossible, thanks in large part to the skill of the program management team at NASA, it should be recognized as one of the most effective public-private partnerships in history."

Elon Musk, SpaceX CEO and CTO,
speaking shortly before the second Falcon 9 flight

On December 8th, 2010, SpaceX became the first commercial company in history to re-enter a spacecraft from Earth orbit. The second Falcon 9 flight – the first under the COTS program – began from Launch Complex 40 at the Cape Canaveral Air Force Station in Florida and, after following a nominal flight profile that included a roughly 9.5-minute ascent, ended two orbits later with the Dragon spacecraft re-entering Earth's atmosphere a few hours later, landing less than one mile from the center of the landing zone in the Pacific Ocean.

Falcon 9, Flight #2 Launch Timeline

Countdown
T–02:35:00 Chief Engineer polled stations. Countdown master auto-sequence proceeded with LOX load, RP-1 fuel load, and vehicle release
T–01:40:00 Master auto-sequence proceeded into lowering the strong-back
T–00:60:00 Master auto-sequence proceeded with second-stage fuel bleed, second-stage thrust vector control bleed
Verification of all sub-auto sequences in the countdown master auto-sequence, except for terminal count
T–00:13:00 SpaceX Launch Director polled readiness for launch
T–00:11:00 Logical hold point

Terminal Count (begins at T–10 minutes)
T–00:09:43 Pre-valves opened to the nine first-stage engines to begin chilling Merlin engine pumps
T–00:06:17 Command flight computer entered alignment state
T–00:05:00 Loading of GN2 into ACS bottle on second stage ceased
T–00:04:46 Internal power on first stage and second stage transferred
T–00:03:11 Arming of flight termination system began
T–00:03:02 Terminated LOX propellant topping and cycled fuel trim valves
T–00:03:00 Verified movement on second-stage thrust vector control actuators
T–00:02:30 SpaceX Launch Director verified – GO
T–00:02:00 Range Control Officer (Air Force) verified range – GO
T–00:01:35 Terminated helium loading
T–00:01:00 Commanded flight computer state to start-up
T–00:01:00 Turned on pad deck and Niagara water
T–00:00:50 Flight computer commanded thrust vector control actuator checks on first stage

T–00:00:40 Pressurized first-stage and second-stage propellant tanks
T–00:00:03 Engine controller commanded engine ignition sequence to start

T–00:00:00 Lift-off
T+0:02:58 First-stage shutdown (main engine cut-off)
T+0:03:02 First stage separated
T+0:03:09 Second-stage engine started
T+0:09:00 Second-stage engine cut-off
T+0:09:35 Dragon separated from Falcon 9 and initialized propulsion
T+0:13:00 On-orbit operations
T+2:32:00 De-orbit burn began
T+2:38:00 De-orbit burn ended
T+2:58:00 Re-entry phase began (entry interface)
T+3:09:00 Drogue chute deployed
T+3:10:00 Main chute deployed
T+3:19:00 Water landing

Before the second Falcon 9 flight, the act of recovering a spacecraft re-entering from Earth orbit was a feat that had previously been achieved by only six nations or government agencies: the US, Russia, China, Japan, India, and the European Space Agency (ESA). In addition to achieving Dragon's first-ever on-orbit performance, eight free-flying payloads were successfully deployed, including a US Army nanosatellite that was the first Army-built satellite to fly in 50 years. As with its maiden flight, the Falcon 9 rocket performed nominally during ascent and staging; after separation of Dragon, the second-stage Merlin engine restarted, carrying the second stage to an altitude of 11,000 kilometers, confirming SpaceX's capability of achieving GTO missions. As Dragon passed over Hawaii, SpaceX received video sent from Dragon on orbit. Then, in preparation for re-entry, Draco thrusters, each capable of producing about 90 pounds of thrust, began the six-minute de-orbit burn. All thrusters performed nominally, although two quads could have been lost and Dragon could still have returned to Earth. During re-entry, Dragon's Phenolic Impregnated Carbon Ablator (PICA-X)[1] heat shield protected the spacecraft from temperatures in excess of 1,650°C. At about 3,000 meters, Dragon's three main parachutes deployed, slowing the spacecraft's descent, demonstrating the safe landing capability that will be required for manned flights. In common with the level of redundancy of many of the vehicle's systems, Dragon could have lost one of its main parachutes and the two remaining chutes would still have ensured a safe landing.

---

[1]   SpaceX had worked with NASA to develop the 3.6-meter PICA-X, a variant of NASA's PICA heat shield. Designed, developed, and qualified in under four years at a fraction of the cost NASA had budgeted for the effort, it is the most advanced heat shield ever to fly and was designed to be used hundreds of times for re-entry with only minor degradation each time.

COTS 2/3

Given the success of the second Falcon 9 flight, it wasn't surprising when SpaceX proposed the possibility of combining two scheduled COTS flight demonstrations of the Falcon 9/Dragon combination. It was planned that the combined flight would precede routine resupply runs to the station under a separate US$1.6 billion fixed-price contract with NASA. To facilitate the previously unplanned ground tests that would be needed to support the combined demonstration flight, the agency boosted its investment in SpaceX by US$128 million in 2011. The combined flight made sense because, at the time, SpaceX was more than two years behind in completing its COTS demonstration flights. In the original plan, SpaceX's second COTS demo had been a five-day mission during which Dragon would approach to within 10 kilometers of the ISS and use its radio cross-link to allow the station's crew to receive telemetry from the capsule and send commands. In the third and final COTS demo, Dragon would berth with the ISS for the first time. Before a decision to merge the two COTS demo missions could be approved by NASA, SpaceX needed to conduct a full-up thermal vacuum test and electromagnetic interference test of Dragon in addition to completing a closed-loop demonstration of the capsule's proximity-operations sensors. They also needed to upgrade its production operation in Hawthorne, add engine test stands at its facility in McGregor, Texas, and make improvements to its launch pad at the Cape. Fortunately, SpaceX (and fellow COTS provider Orbital) had earned a combined US$80 million in milestone payments since the beginning of the 2010 fiscal year – money that came in addition to the payouts negotiated in their original COTS agreements. Also, while both companies were running behind in meeting milestone objectives, they had received the bulk of the funding anticipated under their original COTS contracts, which totaled US$448 million combined.

The extra tests were necessary due to the length of the proposed merged mission. Whereas Dragon had spent just a few hours in orbit on its maiden flight, the merged mission would extend time on orbit to several days. This extra time in LEO would expose Dragon to several long cold-soak and long hot cycles – a thermal stress that hadn't been tested during the first Dragon flight.

In addition to the thermal tests, SpaceX had to fine-tune the rendezvous and proximity ops software that would be required for Dragon's berthing with the ISS. These extra tasks required an improvement in the throughput at SpaceX's facilities – work that NASA, as an investor, was only happy to support by paying SpaceX another US$10 million COTS payout for developing a plan for improvements to its production, test, and launch facilities.

Meanwhile, Orbital, a SpaceX competitor in the private space taxi race, which was developing the Taurus 2 rocket and Cygnus spacecraft under its US$170 million original COTS agreement, also received an extra US$40 million in new milestone payments to help the company prepare for an additional test flight of the Taurus 2, which would loft a dummy Cygnus capsule. Under Orbital's original COTS agreement, the company had been slated to conduct a single test flight of Taurus 2 that would deliver a cargo-laden Cygnus capsule to the ISS, but Orbital decided to bump the mission to the end of 2011 to make room for the additional flight demo.

With Space X and Orbital each planning to launch cargo to the ISS, 2011 was shaping up to be another notable year in the world of commercial spaceflight, with both companies working hard towards realizing their ambition of ultimately ferrying astronauts to the ISS. Unfortunately, the year didn't proceed according to plan for either company.

First there was the nasty business of SpaceX going to court after an industry consultant allegedly spread rumors that SpaceX's rockets were unsafe. In June 2011, SpaceX filed with the Fairfax County circuit court in Virginia, alleging that Joseph Fragola, vice president at tech consulting firm Valador, tried to obtain a hefty deal worth as much as US$1 million from SpaceX at the beginning of June. Fragola claimed that SpaceX needed an independent analysis of Falcon 9 to bolster its reputation with NASA. Not surprisingly, SpaceX didn't take kindly to Fragola's claims, especially when the company learned that the Valador vice president had been contacting officials in the US government to make negative remarks about SpaceX, which created the perception that Fragola claimed SpaceX needed his help to rectify. For example, one of the e-mails that Fragola wrote to Bryan O'Connor, a NASA official at NASA's headquarters, stated that he had heard a rumor that the second Falcon 9 flight had experienced a double engine failure in the first stage and that the stage had blown up after the first stage separated. SpaceX responded, saying that Fragola's rumor was false, explaining that two of the nine engines in the Falcon's first stage had shut down according to plan 10 seconds before the other seven, and there had been no engine failure. Because of Fragola's statements, SpaceX sued for US$1 million for defamation from Valador.

The Valador case wasn't the first time SpaceX had been subject to industry bad-mouthing. Shortly before Fragola's statements, Loren Thompson, a paid consultant for big aerospace companies, had tried to sow doubt about Falcon 9 and Dragon spacecraft to make it seem that NASA was betting the farm on an unproven company. Such a statement ignored the fact that NASA had diversified its portfolio of players, including Boeing and, of course, Orbital. It also sidestepped the fact that SpaceX had always publicly stated its support for competition in the nascent industry. Thompson also made statements on how SpaceX had missed its schedule, while failing to acknowledge the slips of major government developments such as the short-lived defunct Constellation Program or the fact that the Space Shuttle, about to be retired after 30 years of service, was three years late to the launch pad. Another of Thompson's misplaced claims was that SpaceX was breaking its budget. Of course, this was also untrue, since SpaceX, as a commercial provider under COTS, only received money when the company met performance-based milestones, in contrast to Lockheed Martin (one of Thompson's benefactors), developer of the Orion capsule, which had already cost upwards of US$5 billion and was still many years and billions of dollars from completion.

Shortly after the Fragola legal case, Orbital announced that the inaugural flight of their Taurus 2 rocket would be delayed until December 2011, to allow time for the completion and certification of rocket propellant and pressurization facilities at the vehicle's Wallops Island launch site. The maiden flight would be followed by a second Taurus 2 flight two months later, which would demonstrate its space station cargo vehicle, in which the Cygnus capsule would approach the ISS. The delay stemmed from a June test failure of a Taurus 2 first-stage AJ26 engine which caught fire while being tested due to a fuel line breakage.

While the company announced that neither the delay in the propellant facility nor the engine failure would have a financial impact on Orbital's Taurus 2 program, the test failure was another setback for the company which was under contract to make eight cargo delivery runs to the ISS starting in 2012; 2011 had not started well for the company, following the March failure of its smaller Taurus XL rocket, whose fairing had malfunctioned for the second time. In each case, the principal payloads had been NASA science satellites whose combined cost had been estimated at more than US$600 million (the company later successfully placed the US Defense Department's Operationally Responsive Space-1 satellite into LEO using a Minotaur rocket – a converted ICBM that used a fairing that had been redesigned to account for the two Taurus XL failures).

The month after Orbital announced its flight delay, SpaceX settled its lawsuit against Fragola. In August, SpaceX and Valador Inc. said in a joint statement that both sides had agreed to terminate the lawsuit, although terms of the settlement were not disclosed.

The year ended without any flights by either SpaceX or Orbital. In December, Orbital decided to rebrand its new commercial rocket, changing the booster's name from Taurus 2 to Antares (Figure 5.5), saying the name change was to provide clear differentiation between Antares and the Taurus XL.

The name "Antares", which comes from the supergiant star located in the Scorpius constellation, was picked by Orbital because it is one of the brightest stars, and the company expressed the hope that the Antares rocket would turn out to be one of the brightest stars in the space launch vehicle market. The name was also in keeping with the company's tradition of using Greek-derived celestial names for its launch vehicles. After renaming its rocket, Orbital announced it would fly the vehicle twice in the first half of 2012, before beginning a series of operational Cygnus resupply missions to the ISS later in the year; NASA had awarded Orbital a deal covering eight flights valued at US$1.9 billion to deliver cargo to the station through 2015.

While Orbital was kept busy readying the Antares for the first of its two test flights, SpaceX continued its preparations for the third Falcon 9 flight. The second COTS flight would ferry Dragon to the ISS, where it would perform a series of rendezvous and approach maneuvers before delivering cargo to the orbiting outpost. If successful in its first-of-a-kind mission, SpaceX would collect the remaining payments on the US$396 million contract it had with NASA and then enter into a US$1.6 billion agreement for 11 more flights to the ISS.

The first launch attempt, on May 19th, 2012, resulted in a countdown abort at T–00:00:00.5. Following the abort, SpaceX announced the next attempt would be scheduled three days later at 03:44 EDT on May 23rd, 2012. This second attempt was successful and, three days later, after the gumdrop-shaped capsule had

5.5 Antares on the launch pad. Courtesy: Orbital

performed the requisite rendezvous and approach maneuvers, SpaceX became the first commercial outfit to dock its own cargo capsule at the ISS. "It looks like we got us a Dragon by the tail," said US astronaut Don Pettit, who was operating the Canadian-built robotic arm as it reached out and hooked onto Dragon at 9:56 a.m. (13:56 GMT). Experts immediately hailed the historic flight as a new era for private spaceflight. NASA officials agreed. The space agency had been counting on private craft like Dragon to carry cargo, and eventually crew, to the station in the wake of the Space Shuttle's retirement the previous year, and SpaceX had come through.

While attached to the ISS, Dragon delivered 305 kilograms of food, clothing, and other supplies as well as 122 kilograms of cargo bags, 20 kilograms of science experiments, and 10 kilograms of computer equipment. After spending a few days at the orbiting outpost, Dragon made its way home after detaching from the ISS's robotic arm. Five hours later, Dragon used its thrusters to begin its de-orbit burn 400 kilometers above the Indian Ocean, finally splashing down into the Pacific Ocean several hundred kilometers off the coast of Baja California at 08:42 PDT. The capsule was recovered by boats and brought to the Port of Los Angeles. From launch to the splashdown, SpaceX's Dragon mission had lasted 9 days, 7 hours, and 58 minutes. Dragon had joined Russia's Progress, ESA's Automated Transfer Vehicle (ATV), and the Japan Aerospace Exploration Agency's H-II Transfer Vehicle as regular station suppliers. However, Dragon was the only provider in the international line-up with the capability of returning a significant cargo (Dragon's 5,500-pound down-mass capability dwarfs that of Russia's three-person Soyuz) of research samples and equipment in need of refurbishment to Earth – a critical part of future science activities planned for the six-person orbiting science laboratory.

The successful third Falcon 9/Dragon flight wasn't the only win for SpaceX that week. Just days after Dragon docked with the ISS, it was announced that SpaceX and satellite service provider Intelsat had reached a commercial agreement for the launch of a Falcon Heavy rocket. The Falcon Heavy is SpaceX's entry into the heavy-lift launch vehicle market. As the name suggests, the Heavy is a larger vehicle than Falcon 9, comprising a Falcon 9 with two more Falcon 9 stages strapped on either side (Figure 5.6). Each of the strapped-on rockets has nine engines, which will work together as boosters to lift a heavy payload. SpaceX predicts that the Falcon Heavy will launch twice the payload of the Shuttle at about one-fifteenth of the cost of a Shuttle launch, which equates to an approximately 97% reduction in launch costs compared with the Shuttle!

How could SpaceX reduce launch costs by such a margin? Before answering that question, it's worth clearing up how launch prices are determined. When it comes to calculating the costs of government launches, the actual taxpayer cost can only be guessed at by calculating from the cost-plus contract costs, which are usually for multiple launches from the same customer. Now, if SpaceX has multiple launches on its books, the posted price will obviously be reduced according to the number of launches; more launches equals lower costs. At the time of the Intelsat contract, for example, SpaceX had a launch manifest of over 40 payloads divided between Falcon 9 and Falcon Heavy. This number far exceeded any current government contracts, and SpaceX was adding more flights every month.

5.6 Falcon Heavy. Courtesy: SpaceX

Another factor determining cost is rocket performance. The only rocket in service comparable to the Falcon Heavy is the Delta 4 Heavy (Figure 5.7). While a Falcon Heavy looks similar to a Delta 4 Heavy, its performance is much higher, which means its cost per launch is much lower; a Falcon Heavy can put 53 metric tonnes in orbit compared to the Delta 4 Heavy's 23 metric tonnes – a 230% improvement. More importantly, from a customer perspective, a Falcon Heavy costs only about US$100 million per launch, whereas the Delta 4 Heavy costs US$435 million per launch based on an Air Force contract of US$1.74 billion for four launches.

When it comes to calculating payload costs, the Delta 4 Heavy, with its 23-metric-tonne LEO capability, costs about US$19 million per tonne, or about US$8,600 per pound, compared to the Falcon Heavy's price of about US$850 per pound or US$1.9 million per tonne – almost exactly one-tenth of the current Delta 4 Heavy price. It's a huge price differential – one that often prompts the inevitable question: how can the Falcon outperform the Delta by such a wide margin? The main reasons can be found in the development and design of the Falcon 9: (1) low manufacturing cost, (2) low operational cost, i.e. the low man-hours needed per launch, and (3) high-efficiency performance in flight. The low manufacturing cost is a result of the Falcon Heavy's design, which uses three nearly identical rocket stages – a design strategy that translates into more production of the same units and a reduction in unit cost. For example, SpaceX is building towards producing a Falcon 9 first stage or Falcon Heavy side booster every week and an upper stage every two weeks at their plant (Figure 5.8) in Hawthorne, California. At this rate, if this production schedule

5.7 Delta 4 Heavy. Courtesy: United Launch Alliance

5.8 SpaceX's Hawthorne production plant. Courtesy: SpaceX

is achieved, within a few years, SpaceX will be producing more rockets per year than all the rocket companies on the planet combined.

The next key factor in reducing payload costs to LEO is high flight efficiency (although not always resulting in successful launches) – a goal that SpaceX has achieved in the Falcon 1 and 9 rockets by using a short upper stage which consists of a single Merlin engine to place the payload into orbit; for the Falcon Heavy, SpaceX has discussed the possibility of creating a hydrogen–oxygen upper stage, which could boost the Falcon Heavy payload up to 70 tonnes. In addition to the development of the hydrogen–oxygen upper stage, the Falcon Heavy will also benefit from propellant cross-feeding from the side boosters to the center core. During flight, the Falcon Heavy's two outer stages will pump part of their propellant into the center stage (similar to how the Shuttle's external tank fed its main engines). This means these stages will exhaust their propellant faster, but it also means that the center stage has almost a full load of propellant at separation, where it is already at altitude and at speed. It's a method that will give the Falcon Heavy performance comparable to that of a three-stage rocket. (If cross-feed isn't required for lower-mass missions, it is simply turned off.)

So, the Falcon Heavy will be powerful and cost-effective, but what will it be used for? The good news for the commercial launch industry is that the Falcon Heavy opens the door to much larger payloads thanks to the vehicle's large payload fairing – a capability that can be exploited by launching more than just one communications satellite in a single payload. The Falcon Heavy will also be able to launch heavier

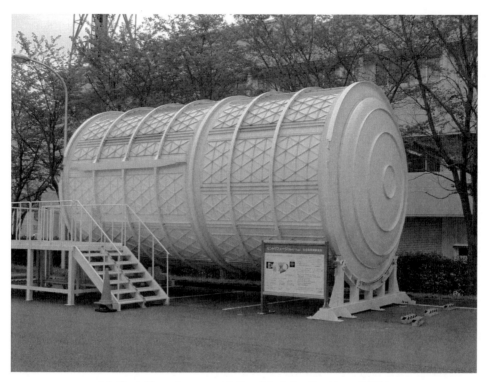

5.9 Centrifuge Accommodation Module. Courtesy: JAXA

payloads, including space station modules such as the Centrifuge Accommodation Module (Figure 5.9), which was the victim of ISS budget-cutting. The Falcon Heavy could also play a role in human missions beyond LEO, perhaps ferrying a long-duration Dragon capsule attached to a crew habitat for an asteroid mission.

Beyond the Falcon Heavy, there is the possibility that SpaceX may build a Falcon Super Heavy (Figure 5.10). Musk has always indicated he intends to continue trying to lower launch costs and improve capability, which means it's unlikely that the Falcon Heavy will be the last vehicle in the SpaceX family of launchers. In fact, the Falcon Heavy, with its 53 metric tonnes of payload, can't be considered a true heavy-lift vehicle (HLV), since these are generally considered to be ones that can lift 70 metric tonnes or more. For example, NASA's planned Space Launch System (SLS; Figure 5.11) is a HLV that will lift 70 metric tonnes in its first version and 130 metric tonnes in a later version.

Based on statements made by Musk (quoted by *Aviation Week*, December 2nd, 2010) and SpaceX president, Gwynne Shotwell (in a letter published in *Space News*, February 7th, 2011), the Falcon Super Heavy will be capable of lifting 150 metric tonnes to LEO at a cost per flight of US$300 million – a rate that will result in a launch cost of US$1,000 per pound, which is close to the per-pound cost of the Falcon Heavy but with a payload three times larger.

While SpaceX basked in the limelight following the success of its third Falcon 9

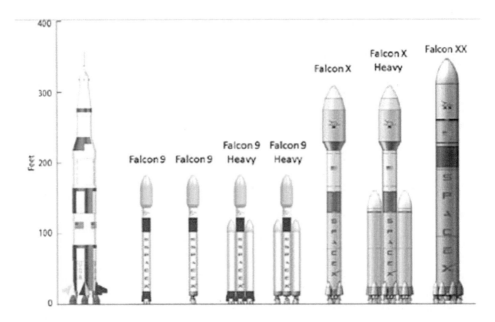

VEHICLE	Falcon 9	Falcon 9	Falcon 9 Heavy	Falcon 9 Heavy	Falcon X	Falcon X Heavy	Falcon XX
1st Stage Engines	Merlin 1D	Merlin 2	Merlin 1D	Merlin 2	Merlin 2	Merlin 2	Merlin 2
Core Diameter (meters)	3.6	3.6	3.6	3.6	6	6	10
Number of Cores	1	1	3	3	1	3	1
Engines per Core	9	1	9	1	3	3	6
Engine Thrust (sea level, lbf)	120k	1.2M	120k	1.2M	1.2M	1.2M	1.7M
Total Lift-off Thrust (lbf)	1.08M	1.2M	3.24M	3.6M	3.6M	10.8M	10.2M
Engine Out Capability?	Yes	No	Yes	No	Yes	Yes	Partial
Mass to LEO (kg)	10.5k	11.5k	32k	34k	38k	125k	140k

5.10 Future SpaceX launch vehicles. Courtesy: SpaceX

flight, the fortunes of Orbital had taken another downturn. A month after Dragon's return, the company announced that its Antares rocket had encountered six or seven weeks of developmental delays and would not be conducting its inaugural demonstration flight until late September or early October, with a second launch carrying the Cygnus cargo vehicle to the ISS planned in mid-December. However, Orbital acknowledged that even this schedule remained tight and could slip further given the program's current development status.

At the time of writing, Orbital was under contract to NASA to deliver cargo to the ISS under two separate contracts. Its COTS contract called for the company to

5.11 Space Launch System. Courtesy: NASA

prove the Antares/Cygnus capability with the demonstration flight and the first Cygnus cargo delivery; this was a cost-sharing contract in which NASA was paying about US$188 million, with Orbital contributing a similar amount. The second NASA contract, called CRS valued at US$1.9 billion, required Orbital to make eight Antares/Cygnus flights to the ISS to deliver a total of 20,000 kilograms of supplies over several years.

Up to the end of 2012, SpaceX had proved its Falcon 9 medium-lift rocket could deliver unmanned payloads to LEO on three occasions. That's not much of a flight record. Yet, since 2009, the company has garnered almost US$1 billion of business in the commercial launch market. That's US$1 billion to a single US company. While it is common in the commercial aviation world to have backlogs, it is unheard of in the space industry. At least it was unheard of until SpaceX came along. Thanks to the company's US$59 million cost per launch, the rest of industry has taken a predictable hit, losing about 40% of the new orders it might have received if SpaceX hadn't been in the game. And, when we're talking about the industry, we're not just talking about the North American market; Europe is being affected too, to the degree that ESA is pushing hard towards meeting a mid-2013 deadline to demonstrate flight hardware ahead of a Société Européenne des Satellites (SES) launch. The reason? SpaceX is contracted to launch the SES-8 satellite into geostationary orbit and, if SpaceX is unable to meet its deadline, the Ariane 5 is the company's backup option. If Space X *does* miss the deadline, it wouldn't be the first time Ariane bails out on one of SpaceX's customers. In 2009, Avanti Communications exchanged a Falcon 9 for an Ariane 5 to launch its K_a-band Hylas 1 broadband satellite because it couldn't wait for SpaceX's launch vehicle to be operational. But, despite lengthy delays and technical issues, Falcon 9 is widely regarded as a commercial success replete with potential. And that potential has already packed a big punch, as it delivered the first of 12 cargo flights to the ISS in October 2012.

6

The Dragon has landed: Picking up where NASA left off

"Looks like we got us a dragon by the tail."
Astronaut Don Pettit as he extended the International
Space Station's robotic arm and captured the Dragon capsule

On May 31st, 2012, SpaceX 's cargo-laden Dragon space capsule parachuted (Figure 6.1) back to Earth after a nearly flawless demonstration mission to the International Space Station (ISS). Splashing down in the Pacific Ocean, about 800 kilometers west of Baja California, the unmanned capsule was picked up by waiting recovery ships

6.1 Dragon following its May 2012 flight to the International Space Station. Courtesy: SpaceX

before being hauled back to the Port of Los Angeles and transported overland to SpaceX's processing facility in McGregor, Texas, where it underwent final inspection.

The landing capped a successful nine-day mission that saw Dragon become the first commercial vehicle to visit the ISS, thereby setting the stage for regular cargo and manned missions to the orbiting outpost – the first of which took place in October 2012. In addition to triggering a series of 12 cargo missions NASA had ordered from SpaceX, Dragon's landmark mission proved that uncrewed cargo vessels sent to the ISS could be recovered and reused and also boosted the company's efforts to make its spacecraft suitable for crewed missions. For SpaceX, a contractor relatively independent of NASA, to have developed, built, tested, and flown its space capsule so successfully, the flight ranked near the top of firsts in the US space program.

Dragon's demo mission also marked NASA's return to flight following the retirement of the Space Shuttle and dispelled some of the doubts that members of Congress had, some of whom had voted to cut Obama's US$830 million budget request for space taxi development. Under NASA's Commercial Resupply Services (CRS) program, SpaceX was contracted to deliver cargo to the ISS beginning in October 2012. The company was also awarded a Commercial Crew Development (CCDev) contract in April 2011 to carry up to seven astronauts, or a combination of personnel and cargo, to and from low Earth orbit (LEO).

Dragon's development, which began in late 2004 when SpaceX started the design of the capsule using its own funding, formed the centerpiece of the proposal SpaceX submitted under NASA's Commercial Orbital Transportation Services (COTS) demonstration program. At 3.6 meters in diameter, Dragon would be smaller than NASA's 5-meter-diameter Crew Exploration Vehicle (CEV) because the SpaceX capsule was intended only for short jaunts to the ISS, and not the longer expeditions to the Moon and beyond.

The gumdrop-shaped Dragon (Figure 6.2) comprises a blunt-cone ballistic capsule, not that dissimilar to the design of the Soyuz or Apollo capsules, a nose-cone cap that jettisons after launch, and a trunk equipped with two solar arrays. To protect its cargo and crew during re-entry, the capsule utilizes a proprietary variant of NASA's phenolic impregnated carbon ablator (PICA) material. The spacecraft also features standard options such as a docking hatch, maneuvering thrusters – 18 of them – and a trunk, which, unlike the rest of the reusable spacecraft, separates from the capsule before re-entry. As shown in Figure 6.3, the Dragon spacecraft sits atop a Falcon 9 booster. During its initial cargo and crew flights, the Dragon capsule will land in the Pacific Ocean and be returned to shore by ship, but eventually SpaceX will install deployable landing gear and use upgraded Super Draco thrusters to permit solid-earth propulsive landings.

Before describing the development and design of Dragon, there may be some who are reading this wondering how the spacecraft got its name. First of all, SpaceX's rule when it comes to naming rockets or rocket parts is that the names must be cool. The Falcon and its siblings were named after the *Millennium Falcon* spacecraft flown by Han Solo in the sci-fi blockbuster *Star Wars* and the Merlin engines may allude to the wizard, Merlin. The business of using cool monikers doesn't stop there; a

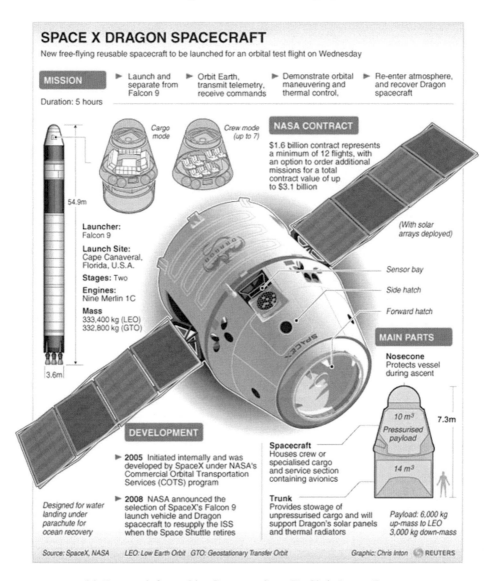

SPACE X DRAGON SPACECRAFT

New free-flying reusable spacecraft to be launched for an orbital test flight on Wednesday

MISSION

Duration: 5 hours

► Launch and separate from Falcon 9
► Orbit Earth, transmit telemetry, receive commands
► Demonstrate orbital maneuvering and thermal control,
► Re-enter atmosphere, and recover Dragon spacecraft

Cargo mode
Crew mode (up to 7)

NASA CONTRACT

$1.6 billion contract represents a minimum of 12 flights, with an option to order additional missions for a total contract value of up to $3.1 billion

54.9m

Launcher: Falcon 9

Launch Site: Cape Canaveral, Florida, U.S.A.

Stages: Two

Engines: Nine Merlin 1C

Mass 333,400 kg (LEO) 332,800 kg (GTO)

3.6m

(With solar arrays deployed)

Sensor bay

Side hatch

Forward hatch

MAIN PARTS

Nosecone
Protects vessel during ascent

10 m³   7.3m

Pressurised payload

14 m³

DEVELOPMENT

► 2005 Initiated internally and was developed by SpaceX under NASA's Commercial Orbital Transportation Services (COTS) program

► 2008 NASA announced the selection of SpaceX's Falcon 9 launch vehicle and Dragon spacecraft to resupply the ISS when the Space Shuttle retires

Designed for water landing under parachute for ocean recovery

Spacecraft
Houses crew or specialised cargo and service section containing avionics

Trunk
Provides stowage of unpressurised cargo and will support Dragon's solar panels and thermal radiators

Payload: 6,000 kg up-mass to LEO 3,000 kg down-mass

Source: SpaceX, NASA   LEO: Low Earth Orbit GTO: Geostationary Transfer Orbit   Graphic: Chris Inton REUTERS

6.2 Dragon infographic. Courtesy: SpaceX, Chris Inton, Reuters

navigation sensor dubbed DragonEye was tested in 2009 on the Space Shuttle *Endeavour*'s STS-127 mission when the Shuttle was approaching the ISS. Then there are the thrusters that Dragon uses to maneuver in orbit; Dragon has 18 of these "Draco" thrusters and, as any Harry Potter fan knows, "Draco" is one of the bad guys in the series of films. The name "Kestrel", given to the engines used to power the upper stage of the Falcon 1 rocket, is no coincidence either, since "kestrel" happens to be the name of a bird in the falcon genus. Dragon meanwhile is named after the fictional creature "Puff the Magic Dragon" in the song by Peter, Paul, and

6.3 Dragon and Falcon 9 in hangar. Courtesy: SpaceX

Mary, because when Musk started his company, many critics reckoned his goals were out of reach. SpaceX's naming trend differs markedly from the one used by NASA, whose space capsule (originally known as the CEV) is known by its acronym MPCV, which stands for Multi-Purpose Crew Vehicle, although the spacecraft sometimes goes by the name of Orion. The agency's creativity in the naming department didn't get much better when it came to thinking up a name for the US$100 billion orbiting outpost, known simply as the International Space Station.

DRAGON DEVELOPMENT

In 2005, the year after SpaceX began developing Dragon, NASA began its COTS development program, soliciting proposals for a commercial resupply spacecraft to replace the soon-to-be-retired Shuttle. SpaceX submitted Dragon as part of its proposal in March 2006 and, six months later, on August 18th, 2006, NASA announced that SpaceX had been chosen, along with Kistler Aerospace, to develop cargo launch services for the ISS. The initial plan called for three Dragon demonstration flights to be flown between 2008 and 2010, for which SpaceX would receive up to US$278 million if they met all NASA's milestones.[1] Two years later, on

---

[1]   Kistler's contract was terminated in 2007 after it failed to meet its obligations and NASA re-awarded Kistler's contract to the Orbital Sciences Corporation.

December 23rd, 2008, NASA awarded a US$1.6 billion CRS contract to SpaceX, with options to increase the contract value to US$3.1 billion. The contract called for 12 cargo flights carrying a minimum of 20,000 kilograms to the ISS.

Dragon's (Figure 6.4) development progressed quickly. On February 23rd, 2009, SpaceX announced that Dragon's heat shield material, PICA-X, had passed heat stress tests in preparation for the capsule's maiden launch. Then, in July 2009, DragonEye, the capsule's primary proximity-operations sensor, was tested during the STS-127 mission, when it was mounted near the docking port of the Space Shuttle *Endeavour* and used while the Shuttle approached the ISS. A later Shuttle mission, STS-129, delivered the COTS UHF Communication Unit (CUCU) and

6.4 The Dragon spacecraft undergoes final preparations at SpaceX's headquarters in Hawthorne, California. Courtesy: SpaceX

6.5 Dragon crew evaluation tests, January 2012. Test crew included (from top left): NASA Crew Survival Engineering Team Lead Dustin Gohmert, NASA Astronaut Tony Antonelli, NASA Astronaut Lee Archambault, SpaceX Mission Operations Engineer Laura Crabtree, SpaceX Thermal Engineer Brenda Hernandez, NASA Astronaut Rex Walheim, and NASA Astronaut Tim Kopra. Courtesy: SpaceX

Crew Command Panel; the CUCU allows the ISS to communicate with Dragon and the Crew Command Panel allows ISS crewmembers to issue basic commands to Dragon. Then, in summer 2009, SpaceX hired former NASA astronaut Ken Bowersox as vice president of their new Astronaut Safety and Mission Assurance Department, in preparation for crews (Figure 6.5) using the spacecraft.

DEMONSTRATION FLIGHTS

Dragon's first flight took place in June 2010 when a stripped-down capsule, dubbed the Dragon Spacecraft Qualification Unit (which was initially used as a ground test bed), was launched on top of a Falcon 9. Dragon's first mission was simply to relay aerodynamic data captured during the ascent; the capsule wasn't designed to survive re-entry because, at the time, SpaceX didn't have a re-entry license (see Appendix II) – a document issued by the Federal Aviation Administration (FAA). Following the successful flight of the first Dragon, SpaceX had to wait for the FAA to issue a re-entry license, which the agency did on November 22nd, 2010 – the first such license ever awarded to a commercial vehicle.

On December 8th, 2010, less than three weeks after being granted the re-entry license, the first Dragon spacecraft launched on COTS Demo Flight 1, which also marked the second flight of Falcon 9. Unlike the qualification unit, this particular Dragon was designed to survive re-entry and was successfully recovered. Following the recovery of the second Dragon, SpaceX continued its preparations for COTS 2, which included further on-orbit testing of the DragonEye sensor, which flew again on STS-133 in February 2011. Then, on April 18th, SpaceX was awarded US$75 million funding during the first phase of NASA's CCDev milestone-based program to help develop its crew system. Under the CCDev program, SpaceX's milestones included the advancement of the Falcon 9/Dragon crew-transportation design, the development and testing of the Launch Abort System (LAS) propulsion design, completion of two crew accommodations demos, full-duration test firings of the launch abort engines, and demonstrations of their throttle capability. At the end of the year, thanks to the success of Falcon 9 and the Dragons, NASA approved SpaceX's decision to combine the COTS 2 and 3 mission objectives (as discussed briefly in Chapter 5) into one Falcon 9/Dragon flight – a flight designated as COTS 2+ or COTS 2/3 (Figure 6.6).

For the COTS 2+, SpaceX used the CRS Dragon variant. This capsule didn't have an independent means of maintaining a breathable atmosphere for astronauts and instead circulated fresh air from the ISS. For typical CRS missions, Dragon will be berthed to the ISS for about 30 days. Capable of transporting 3,310 kilograms of pressurized cargo to the ISS in a volume of 6.8 cubic meters, the CRS Dragon can return 2,500 kilograms of cargo to Earth and the capsule's detachable trunk can transport another 3,310 kilograms of unpressurized cargo in a volume of 14 cubic meters.

DRAGON C2/3 MISSION

Flight Day 1 of the Dragon C2/3 mission began with vehicle power-up and countdown operations that included the final rounds of testing of Falcon 9 and Dragon. This phase included fueling the launch vehicle and configuring Dragon for on-orbit operations. Ten minutes before lift-off, the Terminal Countdown Sequence was initiated during which final configurations of each vehicle were made by computer commands. Three seconds before launch, Falcon 9's nine Merlin 1C engines ignited and, at T–0, the vehicle lifted off (Figure 6.7) from Space Launch Complex 40 at Cape Canaveral Air Force Station, Florida.

After completing a short vertical ascent, Falcon 9 made its now familiar roll-and-pitch maneuver to align itself with the correct flight trajectory. Eighty-four seconds after launch, the vehicle reached maximum dynamic pressure (Max Q) and, three minutes into the mission, the first stage shut down and separated from the vehicle. Seven seconds after stage separation, the second stage's Merlin ignited, commencing a six-minute burn. During the second-stage burn, Dragon's nose cone separated to

6.6 COTS mission patch. Courtesy: SpaceX

6.7 SpaceX's Falcon 9 rocket's nine engines ignite during launch from SpaceX launch pad at Cape Canaveral Air Force Station, May 22nd, 2012. Courtesy: SpaceX

6.8 Dragon with its solar array deployed as it approaches the International Space Station. Courtesy: SpaceX

increase ascent performance, exposing the Spacecraft Docking System (SDS). Nine minutes and 14 seconds after launch, the vehicle shut down its engine and, 35 seconds later, Dragon was released. Shortly thereafter, Dragon deployed its solar arrays (Table 6.1 and Figure 6.8) and SpaceX Dragon Control (Figure 6.9), based in Hawthorne, California, began its list of vehicle status checks to confirm the capsule was in good condition and ready for on-orbit operations.

Table 6.1. Post-launch events.

MET^1	Event
T–0:00:03	Merlin engine ignition
T+0:01:24	Maximum dynamic pressure – Max Q
T+0:03:00	First-stage cut-off
T+0:03:05	Stage separation
T+0:03:12	Second-stage ignition
T+0:03:52	Dragon nose-cone jettison
T+0:09:14	Second-stage cut-off
T+0:09:49	Spacecraft separation

[1] Mission Elapsed Time.

6.9 SpaceX Launch Control. Courtesy: SpaceX

Dragon's first day in space was dedicated to Far Field Phasing Maneuvers[2] and various test objectives (Table 6.2). As the mission progressed, Dragon was required by NASA to complete a series of tests and meet a number of objectives before being permitted to rendezvous with the ISS. One of the first of these tests was to demonstrate its GPS navigation capability – a task that was completed about an hour into the flight. Shortly after demonstrating its way-finding capability, Dragon deployed its guidance, navigation, and control (GNC) bay door to put the necessary rendezvous instruments in place. DragonEye, the vehicle's rendezvous and navigation instrument suite, included a Light Detection and Ranging (LIDAR) imager for this purpose. Just after being deployed, each of these instruments was activated and underwent its requisite checkouts.

---

[2]   The Dragon's rendezvous profile from orbit insertion to docking at the ISS can be divided into three phases: far-field, mid-field, and proximity operations. The far-field stage is characterized as the most quiescent phase. En route to the ISS, the vehicle uses Inertial Measurement Units (IMUs) to calculate its position and uses this opportunity to take extensive ground-based radar updates. The targeting solutions for burn maneuvers are also computed on the ground and uplinked. The mid-field rendezvous phase starts as the Dragon utilizes relative sensors for on-board navigation – a process that is coordinated through timelines anchored to various mission events. The final phase of rendezvous is proximity operations, which involve the frequent use of jet firings to control the relative trajectory of the vehicle up to docking.

Table **6.2**. Initial orbit operations timeline.

MET[1]	Event
00/00:11:53	Solar array deployment
00/00:54:49	Absolute GPS demonstration
00/02:26:48	GNC bay door deployment
00/02:40:49	Relative navigation sensors checkout (LIDAR, thermal)
00/04:11:20	Orbit-adjust burn
00/08:46:52	Full abort test (continuous burn)
00/09:31:25	PCE1 burn
00/09:57:58	Pulsed abort test
00/10:37:58	Free-drift demonstration

[1] Mission Elapsed Time.

On Flight Day 2, Dragon made several engine burns to adjust its orbit in preparation for its rendezvous with the ISS the following day. The burns, referred to as Far Field Phasing or Height-Adjust Burns in NASA parlance, were designed to increase Dragon's orbital altitude. This required the vehicle to target a point 10 kilometers beneath and behind the ISS.

Day 3 maneuvers are depicted in Figure 6.10. Once Dragon had arrived at a point 10 kilometers below and behind the ISS (red arrow), teams at each Mission Control Center conducted a poll before permitting Dragon to conduct its fly-under maneuver to set it up in preparation for performing the R-Bar maneuver (RBM).[3] Once approved, Dragon performed two burns called HA2 and CE2 before entering the ISS's 28-kilometer communications zone in which it could communicate directly with ISS systems. For relative GPS communication with the ISS, Dragon used a proximity communications link supplemented by the COTS Ultra High Frequency (UHF) CUCU aboard the ISS. As it made its close approach, Dragon demonstrated the Relative GPS System (RGPS), which determined the spacecraft's position relative to the ISS. During the test, the RGPS was assessed by comparing data obtained with RGPS with simultaneously acquired absolute GPS data (Figure 6.11). To test the CUCU, ISS astronauts responsible for Dragon operations, Andre Kuipers and Don Pettit, sent a strobe/test command to Dragon. When Dragon

---

[3]   The RBM was a maneuver performed by the Space Shuttle as it rendezvoused with the ISS. The Shuttle performed a back flip that exposed its heat shield to the ISS crew, who took photos of it. Based on analysis of the photos, Mission Control could decide whether the orbiter was safe for re-entry (this was a standard procedure after the *Columbia* accident, caused by a damaged heat shield). The name of the maneuver was based on the R-Bar and V-Bar lines used in the approach of the ISS. R-Bar or Earth Radius Vector is an imaginary line connecting the ISS to the center of Earth. The RBM was developed by NASA engineers Steve Walker, Mark Schrock, and Jessica LoPresti after the Space Shuttle *Columbia* disaster.

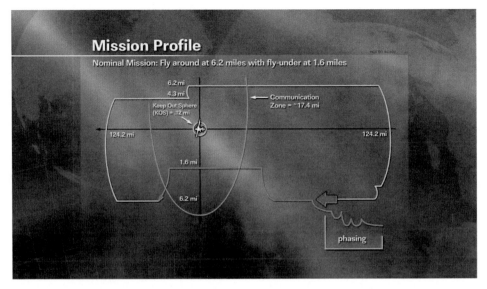

6.10 Dragon's International Space Station maneuvers. Courtesy: SpaceX

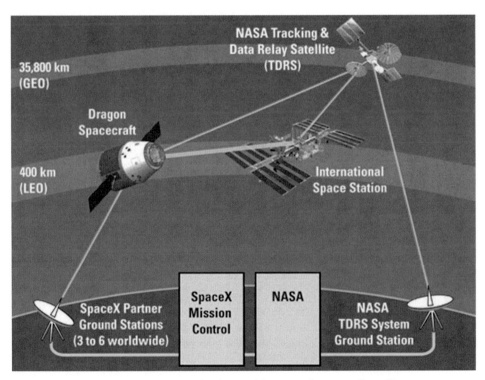

6.11 Dragon communication architecture. Courtesy: SpaceX

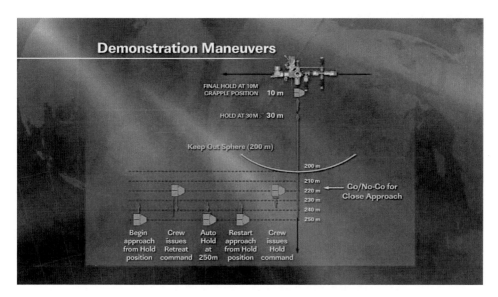

6.12 Dragon fly-around route. Courtesy: SpaceX

received and executed the command, it activated a light for visual confirmation of successful CUCU communications.

Once Dragon had passed 2.5 kilometers below the ISS and all operations were complete, the vehicle made a trajectory adjustment to retreat to a distance of 10 kilometers to begin the ISS fly-around (Figure 6.12). For the fly-around maneuver, the vehicle made several engine burns to cross the ISS's Velocity Vector (V-Bar) flying to a location seven kilometers above the V-Bar before making another maneuver to reduce its velocity. Dragon then passed over the ISS, making one more height-adjust burn to increase the distance between itself and the ISS to 10 kilometers. Dragon once again fired its engines to cross the V-Bar one more time, targeting a point behind and below the ISS. The sequence of events took a full day and finally set the stage for re-rendezvous. The data acquired during the fly-under (Table 6.3) was reviewed by SpaceX and NASA teams, after which the ISS Mission Management Team made a Go/No Go decision for rendezvous.

At the 250-meter hold, several Dragon systems were checked, including rendezvous navigation systems and LIDAR to demonstrate DragonEye's capabilities. Once Mission Control verified that Dragon's position and velocity were accurate, the approach restarted and Dragon made short engine pulses to reinitiate the rendezvous. Once the spacecraft reached 220 meters on the R-Bar, the ISS crew sent a retreat command to demonstrate one of Dragon's rendezvous abort capabilities. Once this had been done, Dragon fired its engines to return to the 250-meter hold point – a maneuver it would be required to perform at any stage during the approach if a retreat command was sent. While Dragon completed the retreat operation, Mission Control assessed whether the spacecraft maintained its range from the ISS and its acceleration and braking performance remained stable.

Table 6.3. Fly-under timeline.

MET	Event
01/23:18	Height-adjustment burn #2
02/00:04	Co-elliptic burn #2
02/00:15	Relative GPS demonstration
02/00:54	Entered ISS communication zone
02/02:44	Crossed R-Bar at 2.5 kilometers
02/03:23	Departure burn #1
02/04:10	Departure burn #2
02/06:47	Forward height-adjust burn #1
02/07:33	Forward height-adjust burn #2
02/12:14	Forward co-elliptic burn #2
03/16:28	Rear height-adjust burn #1

During its maneuvering, Dragon was restricted to the following:
   Closing (axial rate): 0.05 to 0.10 meters per second
   Lateral (radial) rate: 0.04 meters per second
   Pitch/yaw rate: 0.15 degrees per second (vector sum of pitch/yaw rate)
   Roll rate: 0.40 degrees per second
   Lateral (radial) misalignment: 0.11 meters
   Pitch/yaw misalignment: 5 degrees (vector sum of pitch/yaw rate)

Source: International Space Station partnership.

Once all the maneuvers had been verified, Dragon recommenced its approach and demonstrated the second abort scenario, which required the ISS crew to issue a hold command that initiated a period of station-keeping at 220 meters. Once this had been performed, Mission Controllers verified that Dragon's braking performance was nominal and confirmed that the vehicle had stayed within the required range. When all these objectives had been met, Mission Control gave a Go for Close Approach, at which point Dragon fired its engines and closed in on the ISS, entering the Keep Out Zone (Figure 6.13) while the ISS crew watched closely to make sure there were no problems during the approach. As the vehicle approached the 30-meter mark, Dragon made another hold to give the two Mission Control Centers the opportunity to check the vehicle's status and conduct another Go/No Go poll before allowing Dragon to proceed. After being approved to continue its approach, Dragon crept towards the ISS before coming to a stop 10 meters from the station. It had reached the Capture Point. When it was verified Dragon was in the proper position, free drift was initiated and Dragon's thrusters were disabled. The rest of Dragon's rendezvous would be performed by the Canadarm under the control of Don Pettit with assistance from Andre Kuipers. First, the Canadarm captured Dragon (Figure 6.14) and began a carefully choreographed maneuver to place the vehicle above its intended berthing position at the Earth-facing (nadir) Common Berthing Mechanism (CBM) on the Harmony module.

Next, four ready-to-latch indicators were used to confirm that the spacecraft was

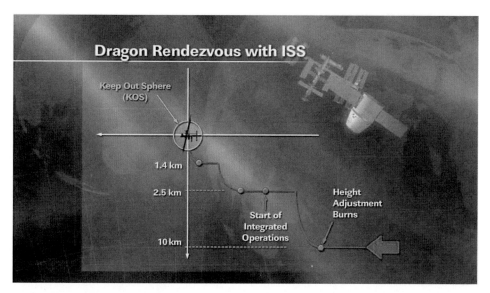

6.13 Keep Out Zone around the International Space Station during Dragon's approach. Courtesy: SpaceX

6.14 Dragon captured by the Canadarm. Courtesy: SpaceX

6.15 Dragon docked with the International Space Station. Courtesy: SpaceX

in the correct position and ready for berthing. Procedures then began to perform first- and second-stage capture of the spacecraft, at the end of which Dragon was secured in place, forming a hard-mate (Figure 6.15) between itself and the ISS. This marked the official start of docked operations. The Canadarm then returned to its pre-grapple position to finish the day's work; the rendezvous up to capture and berthing had taken about eight hours (Table 6.4).

Flight Day 5 began with a new addition berthed to the ISS. One of the first tasks of the day was to pressurize the vestibule between the Dragon and Harmony hatches and check for pressure leaks to make sure the seal between the ISS and the vehicle was tight. Once the leak checks were complete, Mission Control gave the Go to open the hatch, at which point crewmembers began the vestibule outfitting task, which required the installation of ducts and the removal of equipment needed to bolt Dragon in position. Once both hatches were open, the crew conducted air sampling inside Dragon as part of standard ingress operations. Then, Mission Control gave the green flag for Dragon ingress, allowing the crew to commence cargo transfer operations (Figure 6.16).

During the docked phase of the mission, the ISS crew spent about 25 hours conducting cargo operations, which included offloading various items (Table 6.5) and placing them aboard the station. Once this task had been completed, the crew loaded cargo on board Dragon for its return to Earth.

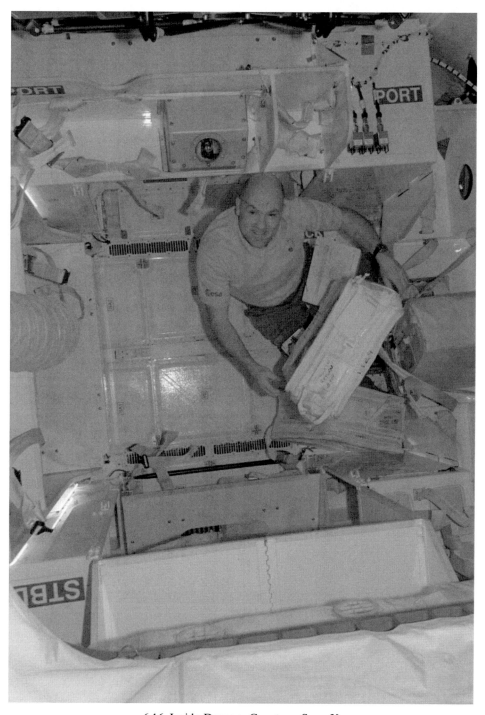

6.16 Inside Dragon. Courtesy: SpaceX

Table 6.4. Rendezvous timeline.

MET	Event
02/18:51	Rear height-adjust burn #2
02/19:37	Read co-elliptic burn #2
02/21:02	Height-adjustment burn #2
02/21:48	Co-elliptic burn #2
02/22:38	Entered ISS communication zone
02/23:16	Height-adjustment burn #3
02/23:32	Mid-course correction #1
02/23:50	Mid-course correction #2
03/00:02	Co-elliptic burn #3
03/00:27	Approach initiation burn
03/00:44	Mid-course correction #3
03/00:59	Mid-course correction #4
03/01:22	R-Bar acquisition – range: 350 meters
03/01:22	180° yaw
03/01:37	Range: 250 meters – station-keeping
03/01:52	Retreat-and-hold demonstration
03/02:17	Range: 220 meters – hold
03/02:32	Entered "Keep Out Zone"
03/03:23	Range: 30 meters – hold
03/03:37	Final approach
03/03:57	Range: 10 meters – Capture Point
03/04:07	Go for Dragon capture
03/04:15	Capture
03/07:36	Berthing

Table 6.5. Dragon C2 cargo manifest.[1]

ISS cargo

Food and crew provisions: 306 kilograms
- 13 bags standard rations: food, about 117 standard meals, and 45 low-sodium meals
- 5 bags low-sodium rations
- Crew clothing
- Pantry items (batteries, etc.)
- SODF and official flight kit

Utilization payloads 21 kilograms
- NanoRacks Module 9 for US National Laboratory: NanoRacks-CubeLabs Module-9 uses a two-cube-unit box for student competition investigations using 15 liquid mixing tube assemblies that function similarly to commercial glow sticks
- Ice bricks: for cooling and transfer of experiment samples

Cargo bags: 123 kilograms
- Cargo bags: reposition of cargo bags for future flights

Computers and supplies: 10 kilograms
- Laptop, batteries, power-supply cables

Total cargo up-mass: 460 kilograms (520 kilograms including packaging)

Table 6.5. *cont.*

Return cargo

Crew preference items: 143 kilograms
- Crew preference items, official flight kit items

Utilization payloads: 93 kilograms
- "Plant signaling" hardware (16 experiment unique equipment assemblies): plant signaling seeks to understand the molecular mechanisms plants use to sense and respond to changes in their environment
- Shear History Extensional Rheology Experiment (SHERE) hardware: SHERE seeks to understand how liquid polymers behave in microgravity by measuring response to straining and stressing
- Materials Science Research Rack (MSRR) sample cartridge assemblies (qty 3): MSRR experiments examined various aspects of alloy materials processing in microgravity
- Other: supporting research hardware such as Combustion Integrated Rack (CIR) and Active Rack Isolation (ARIS) components, double cold bags, MSG tapes

Systems hardware: 345 kilograms
- Multifiltration bed
- Fluids control and pump assembly
- Iodine-compatible water containers
- JAXA multiplexer

EVA hardware: 39 kilograms
- EMU hardware and gloves for previous crewmembers

Total cargo down-mass: 620 kilograms (660 kilograms including packaging)

[1] Courtesy NASA.

Table 6.6. Unberthing timeline.

GMT	Event
04:35	Dragon vestibule outfitting
04:50	IPCU deactivation
05:35	Vestibule depressurization
	Vestibule leak checks (65 minutes)
08:05	Unberthing
09:35	Dragon release

With cargo operations complete, Dragon was closed out, its hatch closed, ducts removed, and control panel assemblies reinstalled. Once Harmony's hatch closed again, the leak check procedure was repeated and the vestibule between the two spacecraft was depressurized to prepare for Dragon's unberthing (Table 6.6). Once again, the Canadarm was used to grapple Dragon before releasing the vehicle. With Dragon flying free, the Canadarm maneuvered the vehicle to its release position 10 meters from the ISS. At this point, Dragon was in Free Drift Mode (FDM) with all thruster systems inhibited. Dragon's navigation instruments underwent the requisite checkouts to ensure the vehicle was receiving correct navigation data and, with all checks complete, both Mission Control Centers gave the Go for release. Then, the

Dragon vehicle was ungrappled, the Canadarm retreated, and Dragon reactivated its thrusters and recovered from FDM. After performing three engine burns to leave the vicinity of the ISS, Mission Control Houston verified that Dragon was on a safe path away from the station.

As it moved away from the ISS, Dragon closed its GNC control bay door to protect the instruments during re-entry. Four hours after release, Dragon was at a safe distance from the station and fired its engines before making the de-orbit burn, taking it on a trajectory to re-enter Earth's atmosphere. Twenty minutes after the de-orbit burn, Dragon hit the entry interface and began to feel the effect of Earth's atmosphere. During re-entry, Dragon's PICA-X heat shield withstood temperatures up to 1,600°C. During the entry phase, Dragon used its Draco thrusters to stabilize its position and control its lift to target the landing. About 10 minutes before splashdown, at an altitude of 13.7 kilometers, Dragon opened its dual drogue chutes, which triggered the main chute-opening command, which occurred at an altitude of three kilometers. Descending under its main chutes, Dragon slowed to its landing speed of 17–20 kilometers per hour and splashed down, landing about 450 kilometers off the California coast.

DRAGONLAB

In addition to the CRS Dragon variant, SpaceX will also fly a version for non-NASA, non-ISS commercial flights – an uncrewed option of the spacecraft known as DragonLab (see Appendix III). Reusable and free-flying, DragonLab (Table 6.7), in

Table 6.7. Specifications of the refurbished Dragon – DragonLab – capsules.

Pressure vessel
- 10 m³ interior pressurized, environmentally controlled, payload volume
- On-board environment: 10–46°C; relative humidity 25~75%; 13.9~14.9 psia air pressure

Unpressurized sensor bay (recoverable payload)
- 0.1 m³ unpressurized payload volume
- Sensor bay hatch opens after orbital insertion to allow sensor access to space and closes upon re-entry

Unpressurized trunk (non-recoverable)
- 14 m³ payload volume in 2.3 m trunk, aft of the pressure vessel heat shield, with optional trunk extension to 4.3 m total length, payload volume increases to 34 cubic meters
- Supports sensors and space apertures up to 3.5 m in diameter

Power, telemetry, and command systems
- Power: twin solar panels provide 1,500 W average, 4,000 W peak, at 28 and 120 VDC
- Communications: commercial standard RS-422 and military standard 1553 serial I/O, plus Ethernet communications for IP-addressable standard payload service
- Command uplink: 300 kbps
- Telemetry/data downlink: 300 Mbit/s standard, fault-tolerant S-band telemetry and video transmitters

common with its CRS sibling, is capable of carrying pressurized and unpressurized payloads with an advertised up-mass of 6,000 kilograms and a down-mass of 3,000 kilograms. Its subsystems include propulsion, power, thermal- and environmental-control, avionics, communications, thermal protection, flight software, guidance and navigation systems, and entry, descent, landing, and recovery gear.

DRAGONRIDER

As the CRS Dragon and DragonLab capsules begin to ferry cargo to and from the ISS, SpaceX is at work developing the manned version of its spacecraft: DragonRider.[4] The development of DragonRider (Figure 6.17) can be traced back to 2006, when Elon Musk announced that SpaceX had built a prototype crew capsule and had tested a 30-man-day life-support system. Later, in 2009, Musk suggested that the crewed Dragon variant was two to three years away from completion. In December 2011, SpaceX performed its first crew accommodations test; the second such test is expected to involve spacesuit simulators and a higher-fidelity crewed Dragon mock-up. This test was followed by a full-duration test of its Super Draco landing/escape rocket engine in January 2012.

When operational, DragonRider will support a crew of seven or a combination of crew and cargo. It will be able to perform autonomous rendezvous and docking with manual override capability and will use the NASA Docking System (NDS) to dock to the ISS. For most missions, DragonRider will remain docked to the ISS for up to 180 days, although, at a pinch, it can extend this period to 210 days – the same as the Russian Soyuz. On the subject of the venerable Soyuz, it's worth mentioning that a DragonRider flight will cost around US$140 million, or US$20 million a seat. That's a third of the cost of the US$63 million cost of a Soyuz flight. As well as being cheaper than the aging Russian spacecraft, DragonRider will probably be safer thanks in part to the integrated pusher launch-escape system that SpaceX will be using. This system has several advantages over the tractor detachable-tower approach used on most prior crewed spacecraft. Not only does the system provide for crew escape all the way to orbit, but it is also reusable and, thanks to the elimination of a stage separation, is safer than traditional escape systems.

---

[4]   Going back to the subject of choosing cool names for their spacecraft, SpaceX's DragonRider moniker may have its origins in the 1997 German children's novel by the same name. Written by Cornelia Funke, *Dragon Rider* follows the exploits of a silver dragon named Firedrake, the Brownie Sorrel, and Ben, a human boy, in their search for the mythical Himalayan mountain range called the Rim of Heaven.

6.17 NASA astronaut Rex Walheim stands inside the Dragon Crew Engineering Model at SpaceX's headquarters in Hawthorne (Los Angeles), California, during a day-long review of the Dragon crew vehicle layout. Courtesy: SpaceX

ONE GIANT LEAP FOR COMMERCIAL SPACEFLIGHT

Dragon's May 2012 flight to the ISS reverberated from Cape Canaveral to Capitol Hill and demonstrated that cargo transport to the ISS can be viably outsourced to commercial players. And, if there were any skeptics after Dragon's return, SpaceX's first CRS mission in October should have removed any doubt that governments now had someone else they could call to send their cargo to LEO.

In August 2012, NASA confirmed SpaceX as the company entrusted with returning Americans to space, when it awarded the company a US$440 million contract with the agency to develop the successor to the Space Shuttle. SpaceX expects DragonRider to be ferrying its first manifest of astronauts some time in 2015, although NASA officials suggest a more cautious timetable, with manned flights commencing some time in the 2016–17 timeframe. Under the Commercial Crew Integrated Capability (CCiCap) program, SpaceX will make the modifications necessary to man-rate DragonRider. Among the modifications is the development of the aforementioned launch-escape system, with powered abort possibilities from launch pad to orbit. SpaceX will also demonstrate that DragonRider will be able to escape a launch-pad emergency by firing integrated SuperDraco engines to carry the spacecraft safely to the ocean. The launch-pad release system that SpaceX has developed keeps the rocket safely on the ground until all first-stage engines are at full power and trending safely. Once released, Falcon 9 can suffer a first-stage engine failure and still make it safely to orbit. Once the launch-pad emergency has been conducted, SpaceX will perform an in-flight abort test that allows Dragon to escape at the moment of maximum aerodynamic drag – an event achieved by firing the SuperDraco thrusters to carry the spacecraft a safe distance from the rocket. Sitting atop the Falcon 9 booster, DragonRider will incorporate a launch-escape system built into the sidewall of the crew vessel itself, allowing for a safe escape path at any point in the flight. Sounds like a safe rocket doesn't it? That's because the launch vehicle was designed from the outset to exceed NASA's crew safety standards: the spacecraft systems are tested to 140% of the maximum expected loads, rather than the 125% that NASA calls for. In contrast, the ATK Liberty vehicle, which is discussed in the next chapter, is based on the 40-year-old solid-rocket architecture which doomed the Space Shuttle *Challenger* and offers none of these safety features. The sobering fact remains that once ignition occurs, solid-rocket boosters – unlike liquid-fueled rockets – can't be turned off. Launch vehicles are twitchy at the best of times and any launch vehicle can have a bad day; the problem with solid boosters is the tendency to turn a bad day into something much, much worse.

In addition to these critical tests, SpaceX is working on a breakthrough propulsive landing system for gentle ground touchdowns on legs and is refining and rigorously testing essential aspects of DragonRider's design such as its life-support system and an advanced cockpit design. All these tests, and more, are needed to demonstrate that all DragonRider's systems meet NASA requirements. Once all the checks and tests have been conducted, the Falcon 9/DragonRider combination, with its minimal number of stage separations, all-liquid-rocket engines that can be throttled and turned off in an emergency, engine-out capability during ascent, and powered abort capability all the way to orbit, will be the safest spacecraft ever developed.

With the selection of DragonRider, NASA made a critical decision regarding a new chapter in American space exploration. The decision was about much more than whose brand name would be emblazoned on the side of a spacecraft; it was, in many ways, a referendum on the merits of a new approach to developing and conducting spaceflight operations. In choosing SpaceX, NASA opted for a path that holds the potential to shape the future of American space exploration, but that decision was far from a foregone conclusion. With so much at stake, many reasoned that SpaceX, a company with only a decade of experience under its belt, simply lacked the history to compete with more experienced aerospace contractors. But many of the naysayers seemed to have forgotten that the alternative of a traditional approach of top-down, sole-source, cost-plus contracting just wasn't viable any longer.

COMMERCIALIZING LEO

December 7th, 2012, marked the 40th anniversary of the flight of Apollo 17. That flight was the last time the US launched an astronaut beyond Earth orbit. The Apollo Program lasted less than four years and one of the reasons the program had such a short lifespan was because the US tied itself to an infrastructure and a way of doing business that was just way too expensive to sustain. It took the agency a long time to realize this. In fact, it wasn't until 2006 (Project Constellation or "Apollo on Steroids") that NASA finally bit the commercial bullet by establishing COTS, a program designed to see whether there was a better way of doing business for achieving access to space by harnessing the innovation and drive of private industry. After investing nearly US$8 billion developing the Ares I launcher, NASA saw the writing on the wall; the Constellation Project was just too expensive to sustain. Instead, it was the COTS approach for cargo delivery to the ISS that the agency adopted for its commercial crew program.

As for SpaceX's experience, don't forget that the company is contractually obligated to 12 flights in the coming years as part of its CRS. So, by the time DragonRider *is* ready to ferry astronauts to the ISS, SpaceX will have flown the route several times using the whole system – Falcon 9, Dragon (albeit the cargo version), ground operations, tracking, space station rendezvous, and berthing. It will be a well-proven, *experienced* system.

While SpaceX's first COTS flight to the ISS took longer than expected, the results were well worth NASA's time and money. On the back of an investment of US$396 million, plus a great deal of advice, NASA made it possible for SpaceX to produce not just a new launch vehicle, but a real game-changer – a commercial space transportation system, all for less than the space agency spent on one suborbital test launch of the Ares I-X booster in 2009. It was also less than NASA spent on the development of its MPCV crew vehicle in the first half of 2012 alone; no matter how you look at the NASA–SpaceX deal, the Falcon–Dragon transportation system probably represents the best investment NASA ever made. Thanks to SpaceX, which has a contract to launch 12 cargo flights to the ISS, not only is the US back in the orbital transport business, but the stunningly low cost of the launch system represents a rare bargain for taxpayers. In short, SpaceX and its Dragon are indisputable proof that a new approach to space transportation can work far more effectively than the old ways.

As I'm writing this, DragonRider was in the news again with the announcement by NASA that the capsule had passed a design review for a manned launch. Tests seem to be on track for a first unmanned launch some time in 2017. The news about Dragon came only a couple weeks after NASA had news of its own with the unveiling of its Orion. Orion is built by the private sector – mostly by Lockheed Martin – and is scheduled to make its first unmanned flight in 2014, with the Space Launch System (the rocket that will launch Orion) scheduled to test launch in 2017. It's great to see NASA on track for a major spaceflight milestone again, even if the capsule may not evoke the same awe as the Space Shuttle. But, compared to the SpaceX CCDev 2 program, the Space Launch System (SLS) that Orion is a part of is expected to cost US$38 billion. That isn't a typo. It's US$38 billion.[5] And between US$17 billion and US$22 billion of that will be spent just for development. That equates to *80 times* the cost of the commercial development of four manned crew vehicles, or about 320 times more per vehicle. Okay, so we know that Orion is being designed for missions to the Moon and perhaps beyond, and getting out of LEO is more complicated than simply reaching LEO, but is it 320 times more complicated? Incidentally, Elon Musk reckoned he could beef up DragonRider and get to the Moon for around US$3 billion, which would be less than one-quarter of the US$13 billion NASA spent on the defunct Constellation Moon program that never produced a flyable rocket.

That's not to say we should do away with the NASA system in favor of CCDev-style programs. After all, Dragon and its variants wouldn't exist if it were not for NASA's help. And, despite the promise of Dragon, SpaceX is still a very new company and has yet to prove itself in the manned spaceflight arena. So, while we all know that NASA has spending problems, it's probably not a good idea to overthrow the current system in favor of an industry that is only now taking its first steps. After all, SpaceX has yet to demonstrate that it can sustain a flight rate that will support their government and commercial commitments *and* do it at the quoted prices. The May 2012 Dragon flight to the ISS demonstrated to the world that SpaceX is a legitimate player in the LEO cargo delivery business and opened up new possibilities and markets in LEO; hopefully, by the end of this decade, or maybe sooner, DragonRider will have ferried dozens of astronauts to orbit and, in so doing, will have taken another huge step.

---

[5] The number was quoted in the *Orlando Sentinel* on August 5th, 2011. The *Sentinel* reported that NASA estimated the cost of the SLS and Orion Multi-Purpose Crew Vehicle (MPCV) could be as much as US$38 billion through 2021. The estimate, which came from an internal NASA report obtained by the *Sentinel*, stated that the cost of developing the SLS and the MPCV through 2017, the date of the first unmanned flight, was US$17 billion to US$22 billion. Getting the vehicles ready for the first manned mission in late 2021 would be an additional US$12 billion to $US16 billion.

7

The space taxi race

"NASA is on a good track to turn over astronaut transportation to commercial operators, and I think ultimately the agency will be successful at doing that."

Orbital CEO Dave Thompson, despite not winning a CCDev 2 award

"NASA announces $1.1 billion in support for a trio of spaceships." That was one of the headlines on August 3rd, 2012, when NASA announced it had committed US$1.1 billion over the next 21 months to support spaceship development efforts by a line-up of companies with the aim of having American astronauts flying on American spacecraft within five years. With SpaceX hawking much of the media spotlight following the spectacular success of Dragon's docking with the International Space Station (ISS), it had been easy to forget there were other major players in the race to build commercial space taxis. So, before continuing with the SpaceX story, it's worth taking a look at some of the other companies developing manned vehicles capable of flying astronauts to and from the orbiting outpost. These companies include aerospace juggernaut, Boeing, who received US$460 million in the Commercial Crew Integrated Capability (CCiCap) outlay, and Sierra Nevada Corporation (SNC), who received US$212.5 million. Also in the space taxi hunt are outliers the Orbital Sciences Corporation (OSC), Blue Origin, and ATK Aerospace Systems.

NASA's CCiCap announcement heralded the next phase of the agency's commercial spaceflight effort. In short, CCiCap calls for Boeing, SNC, and SpaceX to take their design and testing program through a series of milestones by May 2014. Optional milestones could lead to crewed demonstration flights in later years. The goal of the program is to have at least one commercial space taxi ferrying astronauts to and from the ISS by 2017. The three companies say they are confident they can meet or beat that schedule – provided they continue to receive NASA support. This is good news for NASA Administrator, Charles Bolden, who has to fly his astronauts on Russian hardware at the exorbitant price of US$63 million a seat (SpaceX, by comparison, promises seats for around US$20 million). With the US at the mercy of a sole rocket provider, it's no surprise that the space taxi program is a top priority of the Obama Administration.

CCiCap is the third phase of NASA's Commercial Crew Program (CCP). In earlier phases, Boeing, SpaceX, and SNC had received hundreds of millions of dollars in NASA support. While SpaceX is rapidly upgrading its Dragon capsule to manned capability, Boeing is working on its CST-100 and SNC is testing its Dream Chaser spaceplane, which looks like a miniature version of the Space Shuttle. Leading up to the August 2012 decision, NASA and congressional leaders made a deal that called for two commercial partners to receive full funding, with one backup partner receiving half funding. Looking at the numbers, it would seem that SNC drew the short financial straw, since the company's milestones stopped just short of a critical design review, while SpaceX and Boeing could be funded through that phase, but none of the companies was complaining. In a statement, Elon Musk, SpaceX's CEO and chief designer, hailed the CCiCap award as "a decisive milestone in human spaceflight" that would set "an exciting course for the next phase of American space exploration". Boeing's statement struck a similar tone, with John Elbon, Boeing vice president and general manager of space exploration, announcing:

"Today's award demonstrates NASA's confidence in Boeing's approach to provide commercial crew transportation services for the ISS. It is essential for the ISS and the nation that we have adequate funding to move at a rapid pace toward operations so the United States does not continue its dependence on a single system for human access to the ISS."

The CCiCap funding was partly thanks to the success of SpaceX because, before the Dragon flight, the plan for supporting numerous competitors had been in danger. Some in Congress, such as NASA Appropriations Chairman Frank Wolf of Virginia, had been pressuring the agency to select a single provider immediately to save money. But, following Dragon's successful ISS flight, Wolf was persuaded to let the competition continue. This, for NASA, was a good thing, because the agency didn't want to be dependent on a single means of getting its astronauts into orbit. They had tried this strategy with the Space Shuttle and it hadn't worked out; the Space Shuttle program was down for more than five years total during its lifetime for investigations after accidents.

The companies not awarded contracts were not out of the space taxi competition; it just meant they would have to keep building using their own money and no taxpayer assistance. For now. Whether or not they were willing to do so depended on how deep their pockets were, what they thought the prospects of getting back in the game were, and what markets they foresaw beyond NASA ISS crew support. After all, the ISS won't be the only low Earth orbit (LEO) destination for long; Bigelow Aerospace, constructor of inflatable habitats, plans private space facilities, for which they have memorandums of understanding with several sovereign clients, including Japan and the Netherlands.

Whichever company or companies NASA selects for carrying its crew, the new vehicle will go beyond a Shuttle replacement in one important way. Since the ISS was first occupied more than a decade ago, there has always been a Soyuz capsule (Figure 7.1) berthed to provide a rescue capability. The reliance on the Russians for the lifeboat service was because the Space Shuttle never had the ability to stay at the

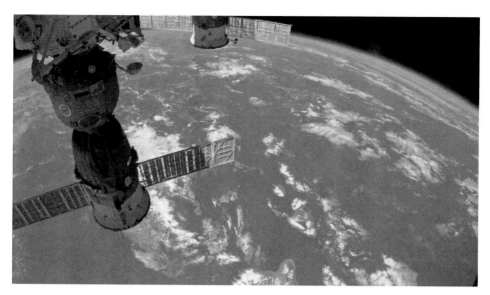

7.1 Soyuz docked with the International Space Station. Courtesy: NASA

station for more than a week or so. However, all the new vehicles are designed for an orbital life of several months and, with room for up to seven crewmembers, they are larger than the cramped confines of the three-passenger Soyuz. Using one or more of these commercial capsules as the ISS's lifeboat service will be a game-changer, not only because it will eliminate NASA's dependence on the Russians (with whom relations can be rocky), but also because these new spacecraft will allow an expansion of ISS crew capacity, currently limited by lifeboat size.

This chapter profiles some of the companies in the space taxi race – a competition that wouldn't have been possible without NASA's Commercial Crew and Cargo Program. The program was established to invest financial and technical resources to catalyze efforts within the commercial sector to develop and demonstrate safe, reliable, and cost-effective space transportation capabilities. The program manages Commercial Orbital Transportation Services (COTS) partnership agreements with US industry totaling US$800 million for commercial cargo transportation demonstrations. It also oversees the Commercial Crew Development (CCDev), a NASA investment funded by US$50 million of the American Recovery and Reinvestment Act (ARRA) funds, to stimulate efforts within the private sector that aid in the development and demonstration of safe, reliable, and cost-effective space transportation capabilities. CCDev 2 commercial partners include Blue Origin, Boeing, SNC, and, of course, SpaceX. In essence, the Commercial Crew and Cargo Program is a multiphase strategy that doles out funds to companies to develop solutions for crew transportation to LEO.

The August 2012 CCiCap announcement, which was the final-phase development funding under NASA's CCP, established Boeing and SpaceX as the clear frontrunners in the space taxi race, with SNC waiting in the wings as the fallback system in case one of the other two falters. Passed over for a CCiCap award was ATK Aerospace, the long-time supplier of solid-rocket motors for the now-retired Space Shuttle. ATK, whose Liberty design had not been funded in previous rounds of the CCP, would have used an ATK-built solid-fuel core stage and the first stage of Europe's Ariane 5 rocket as an upper stage. The company might still build its Liberty system by continuing to work through an unfunded Space Act Agreement (SAA). Meanwhile, if Boeing and SpaceX meet their NASA-approved milestones in the 21-month CCiCap performance period, their designs will undergo a critical design review (CDR). If the CDRs go well, construction can begin. The three CCiCap winners said they could stage their first unmanned demonstration flights by 2015 or 2016, although construction and flight tests are not funded under the initial CCiCap awards. Boeing officials said the CST-100 could be ready to conduct its first manned flight by 2016.

And what about Blue Origin? The secretive space start-up bankrolled by Amazon.com founder Jeff Bezos didn't submit a CCiCap proposal, although the company had been involved in NASA's CCP since the first round of funding was awarded in 2010. The Kent, Washington, company had received US$25.7 million in NASA CCDev 2 funding, some of which it had put towards a crew escape system for its New Shepard vertical-take-off, vertical-landing (VTVL) suborbital vehicle.

In its assessment of the proposals, NASA showed a clear preference for complete transportation systems rather than proposed subsystems such as those put forward by companies such as Paragon Space Development Corporation, which won one of the five first-round CCDev awards to work on a life-support system. Other companies were cut during the assessment for having major weaknesses or for simply failing to follow the instructions. The initial evaluation of the eight finalists – ATK, Blue Origin, Boeing, Excalibur Almaz, OSC, SNC, SpaceX, and United Launch Alliance (ULA) – were re-evaluated based on technical and business criteria and given a color-coded score for each, although these ratings only served as indicators and didn't form the only basis of the final decision. What NASA was looking for was a diverse range of technical approaches – a strong business approach and spacecraft-development versus launch vehicle development. This last requirement made sense, since the US has plenty of launch vehicle development expertise and experience, but very little skill developing crew-carrying spacecraft.

When it came to selecting the winning three, the Boeing and SpaceX proposals stood out from the rest thanks to their high ratings in technical and business factors. It also helped that the two companies were developing capsules. Since NASA considered it important to have at least one lifting-body concept in the portfolio, the race for the final funding dollars was between OSC and SNC. SNC, which scored high in business organization and demonstrated a strong commitment to the public–private partnership associated with the CCP, also edged out Orbital in crew-carrying capability: seven versus four.

BOEING

Profile
Spacecraft: CST-100
Type: Capsule with service module
Crew capacity: 7
Launch vehicle: Atlas V (United Launch Alliance)
CCiCAP funding: US$460 million
CCiCAP term: 21 months
Previous CCDev funding (including optional milestones): US$130.9 million (Boeing), US$6.7 million (ULA)
Total CCDev and CCiCAP funding (if all milestones met): US$590.6 million (Boeing), US$6.7 million (ULA)

Rather than using heritage technology from the lifting-body program, Boeing is advancing plans for its new capsule-based spaceship. The CST-100 capsule (Figure 7.2) is a spacecraft design proposed by Boeing in collaboration with Bigelow Aerospace as their entry for the CCDev program. As with all the other competitors in the space taxi race, the CST-100's primary mission is transporting crew to the ISS and to private space stations such as the proposed Bigelow Aerospace Commercial Space Station. At first glance, the gumdrop-shaped capsule looks similar to the Apollo and Orion (Figures 7.3 and 7.4), the latter a spacecraft being built for NASA by Lockheed Martin.

When complete, the CST-100 (incidentally, the number "100" stands for 100 kilometers, the height of the Kármán line, which defines the boundary of space) will be larger than the Apollo command module but smaller than Orion. Capable of ferrying crews of up to seven as the result of a generous habitable interior and the reduced weight of equipment needed to support an exclusively LEO configuration, the CST-100 is designed to remain on orbit for up to seven months and for reusability for up to 10 missions. Although the initial launch vehicle for the CST-100 will most likely be the Atlas V, the spacecraft will be compatible with multiple launch vehicles, including the Delta IV and Falcon 9.

In common with SNC, Boeing is also the recipient of earlier NASA funding. In the first phase of the CCDev program, the company was awarded US$18 million for preliminary development of the spacecraft – a sum that was followed by another US$93 million in the second phase for further development. Although an industry juggernaut, Boeing was still reliant on external funding for development of the CST-100, the company stating in July 2010 that the capsule could only be operational in 2015 with sufficient near-term approvals and funding. Boeing also indicated they would proceed with development of the CST-100 only if NASA implemented the commercial crew-transport initiative that was announced by the Obama Administration in its FY11 budget request. In fact, Boeing's business case was reliant not only on continued NASA funding, but also on the existence of a second destination, hence the partnership with Bigelow.

While the CST-100 is sometimes confused with its big-budget cousin, Orion, the

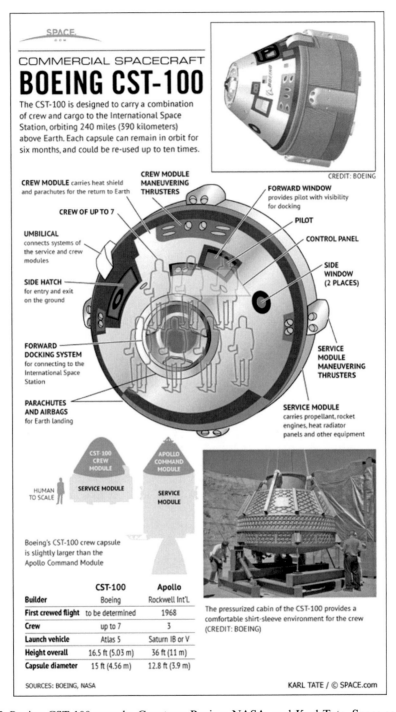

7.2 Boeing CST-100 capsule. Courtesy: Boeing, NASA, and Karl Tate, Space.com

7.3 Orion capsule. Courtesy: NASA and Boeing

7.4 Orion capsule airbag landing. Courtesy: NASA

Boeing vehicle has no Orion heritage, its design drawing mostly upon Boeing's experience with Apollo, the Shuttle and ISS programs, and the Department of Defense's Orbital Express project.

Measuring 4.5 meters across at its widest point and standing 5.03 meters high (with service module attached), the CST-100 sports a four-engine "pusher abort system" rather than an escape tower, as used in the Mercury and Apollo programs, and is designed to land on terra firma, although it can also support a water landing in the event of an abort. For docking with the ISS, the vehicle will use the Androgynous Peripheral Attach System (APAS) while for re-entry, the Boeing Lightweight Ablator (BLA) heat shield will protect the crew.

Work on CST-100 design, manufacture, testing, and evaluation is well underway and moving at a rapid pace thanks to Boeing pulling in proven technology. For example, the CST-100 utilizes technology developed to build the Apollo-era heat shield and the Space Shuttle's thermal protection system, as well as the autonomous rendezvous and docking gear developed on the Pentagon's experimental satellite-refueling Orbital Express mission. The CST-100 also uses flight computers currently operating on the Boeing-built X-37 spaceplane (Figure 7.5). Given that Boeing is working on a fixed-price development, using mature designs and drawing upon flight-proven hardware not only makes sense financially, but also gives them an edge on the competition.

Some of the more notable tests performed by Boeing in its development of the CST-100 include drop tests to validate the design of the airbag cushioning system and the capsules' parachute system. The airbags, which are deployed by filling with a mixture of compressed nitrogen and oxygen, are located underneath the CST-100's heat shield, which is designed to be separated from the capsule while under parachute descent at about 1,500-meter altitude. In September 2011, Boeing conducted drop tests in the Mojave Desert in California, at ground speeds between 16 and 48 kilometers per hour to simulate cross-wind landing conditions. Bigelow Aerospace, one of Boeing's partners in the CST-100, built the mobile test rig and conducted the tests. The drop tests were followed up by parachute tests in April 2012, when Boeing dropped a CST-100 mock-up over the Nevada desert at Delamar Dry Lake near Alamo, Nevada, successfully testing the craft's three main landing parachutes from 3.35 kilometers. Bigelow Aerospace President, Robert Bigelow, was impressed, noting in a press statement:

> "If astronauts had been in the capsule during these drop tests, they would have enjoyed a safe, smooth ride ... further proof that the commercial crew initiative represents the most expeditious, safest, and affordable means of getting America flying in space again."

As this book is being written, Boeing is preparing for a landing airbag test series, a forward heat-shield jettison test, and an orbital maneuvering/attitude control engine hot-fire test, all of which will help engineers gather more data on design aspects of the CST-100.

Meanwhile, manufacturing of the CST-100 continues at the Orbiter Processing Facility-3 at Kennedy Space Center – a facility that was leased to Boeing through a

7.5 X-37. Courtesy: NASA and USAF

partnership with Space Florida in October 2011. If all goes well, Boeing's private spaceship may be ready to carry astronauts to the ISS, Bigelow habitats, and other LEO locales by 2016. Casting an eye towards the business case, Boeing expects markets to materialize but, given the unpredictability of the private commercial spaceflight market, the company is hedging its bets by having a low, medium, and high business model. Those business models consider existing partnerships such as the one Boeing has with Bigelow Aerospace, which is developing large and leasable expandable space habitats, and a partnership with space tourism provider Space Adventures, which intends to sell unused seats on the CST-100 for flights to LEO.

Boeing has 19 milestones (Table 7.1) to meet during the 21-month CCiCAP program – more than SpaceX (14) and SNC (10). The goal is to take the CST-100 capsule and its ULA-supplied Atlas V rocket through a CDR in April 2014. If Boeing completes all milestones, NASA will pay it US$460 million. The company also has 35 optional milestones in its plan that could earn Boeing additional awards if the company completes them and NASA has funding available (during the CCDev 2 funding round, the company earned an additional US$20.6 million in addition to the US$92.3 million the company was awarded for achieving its base milestones).

Table 7.1. Boeing CCiCAP base milestones.[1]

No.	Description	Date	Amount
1	For the Integrated System Review (ISR), Boeing has to establish and demonstrate a baseline design of the Commercial Crew Transportation System (CCTS) integrated vehicle and operations that meets system requirements	August 2012	US$50 million
2	For the Production Design Review (PDR), Boeing has to establish the baseline plan, equipment, and infrastructure for the CST-100's manufacture, assembly, and acceptance testing	October 2012	US$51.7 million
3	For the Safety Review Board (SRB), Boeing has to prepare and conduct a Phase 1 Safety Review of Production Design Review-level requirements, vehicle architecture, and design to assess conformance with NASA's Crew Transportation System (CTS) certification process	November 2012	US$25.2 million
4	For the Software Integrated Engineering Release, Boeing has to demonstrate the software for the guidance, navigation, and control (GN&C) system for the flight ascent phase	January 2013	US$20.4 million

No.	Description	Date	Amount
5	For the Landing & Recovery/Ground Communication Design Review (GCDR), Boeing has to conduct a Landing & Recovery/GCDR which establishes the baseline plan for equipment for conducting CST-100 flight operations fulfilling ground communications and landing/recovery operations	January 2013	US$28.8 million
6	The Launch Vehicle Adapter (LVA) Preliminary Design Review (PDR) demonstrates that the preliminary design meets requirements with acceptable risk and within cost and schedule constraints and establishes the basis for proceeding with detailed design	February 2013	US$45.5 million
7	For the Integrated Stack Force and Moment Wind Tunnel Test, Boeing has to develop a test matrix, fabricate the necessary test models, and perform an integrated launch vehicle force and moment wind tunnel test to validate predictions on integrated Crew Module (CM)/Service Module (SM)/Launch Vehicle (LV) stack for ascent	April 2013	US$37.8 million
8	The Dual Engine Centaur (DEC) Liquid Oxygen Duct Development Test is self-explanatory	May 2013	US$21.5 million
9	For the Orbital Manoeuvring and Attitude Control (OMAC) Engine Development Test, Boeing has to complete this test to support component, subsystem, and CST-100 vehicle-level development	July 2013	US$50.2 million
10	The Spacecraft Primary Structures Critical Design Review (CDR) confirms that the requirements, detailed designs, and plans for test and evaluation form a satisfactory basis for fabrication, assembly, and structural testing	October 2013	US$8.6 million
11	The Service Module Propulsion System Critical Design Review test will be conducted after major SM propulsion components have completed their individual CDR, which confirms the requirements, detailed designs, and plans for test and evaluation form a satisfactory basis for production and integration	November 2013	US$7.5 million

Table 7.1 *cont.*

No.	Description	Date	Amount
12	The Mission Control Center Interface Demonstration Test demonstrates the linkage between the MCC and the Boeing Avionics Software Integration Facility, which is a precursor to integrated simulation capability for flight operations training	September 2013	US$7.9 million
13	The Launch Vehicle Adapter Critical Design Review confirms that the requirements, detailed designs, and plans for test and evaluation form a satisfactory basis for production and integration	September 2013	US$13.5 million
14	The Emergency Detection System (EDS) Standalone Testing is self-explanatory	October 2013	US$13.8 million
15	The Certification Plan Review defines Boeing's strategy leading to a crewed flight test	November 2013	US$5.8 million
16	The Avionics Software Integration Lab (ASIL) Multi-String Demonstration Test demonstrates the flight software closed loop with GN&C for the flight ascent phase	December 2013	US$24.9 million
17	The Pilot-in-the-loop Demonstration ensures all key hardware/software interfaces for Manual Flight Control meet requirements, including operational scenarios and failure modes	February 2014	US$13.9 million
18	The Software Critical Design Review confirms that the requirements, detailed designs, and plans for test and evaluation form a satisfactory basis for flight software development, verification, and delivery	March 2014	US$15.1 million
19	The Critical Design Review (CDR) Board establishes and demonstrates a critical baseline design of the CCTS that meets system requirements	April 2014	US$17.9 million
Total:			*US$468 million*

[1] Boeing passed the first of the CCiCap milestones – the ISR – at a three-day meeting in August 2012, which covered the seven-seat capsule and Atlas V and mission operations on the ground and in space.

At the time of writing, the CST-100 program is on track to deliver its first flight-design hardware – the one-piece lower section of the capsule's aluminum pressure vessel – in less than 20 months, with a first flight tentatively scheduled by the end of 2016, although Boeing is looking at ways to launch earlier.

SIERRA NEVADA CORPORATION

Profile
Location: Louisville, Colorado
Spacecraft: Dream Chaser
Type: Lifting-body
Crew capacity: 7
Launch vehicle: Atlas V (United Launch Alliance)
CCiCAP funding (if all milestones met): US$212.5 million
CCiCAP term: 21 months
Previous CCDev funding: US$125.6 million (Sierra Nevada), US$6.7 million
   (ULA)
Total CCDev and CCiCAP funding (if all milestones met): US$338.1 million
   (Sierra Nevada), US$6.7 million (ULA)

SNC's private spaceplane aims to pick up where the Space Shuttle left off. At least that's the impression one gets when viewing the vehicle (Figure 7.6) that could easily pass for a miniature Shuttle. But, while the spacecraft's design certainly takes cues from the past, the company is hoping its winged spaceship – Dream Chaser – will redefine commercial spaceflight in much the same way as the now-retired Space

7.6 Dream Chaser. Courtesy: Sierra Nevada Corporation

7.7 Dream Chaser attached to Atlas V 402. Courtesy: Sierra Nevada Corporation

Shuttle changed manned spaceflight in the 1980s. Headquartered in Louisville, Colorado, SNC's Space Systems division designs and manufactures spacecraft, rocket motors, and spacecraft subsystems for the US government and commercial customers. While not in the same category as Boeing, SNC Space Systems has more than 400 successful space missions under its belt, delivered more than 4,000 systems, subsystems, and components, and has concluded over 70 programs for NASA.

Dream Chaser is a crewed suborbital/orbital (the vehicle was originally planned in 2004 to be a suborbital vehicle modeled after the X-34 but the design was revised in 2005 and is now based on NASA's HL-20 lifting-body design), vertical-take-off, horizontal-landing (VTHL) lifting-body spaceplane that will carry up to seven crew to and from LEO. Designed to launch vertically on a man-rated Atlas V 402 rocket (Figure 7.7) and land horizontally on conventional runways, the primary mission of the reusable composite Dream Chaser is to provide NASA with a safe and reliable commercially operated transportation service for crew and cargo to and from the ISS. Future missions may include delivering crew and cargo to other orbiting facilities such as Bigelow's habitats, functioning as a short-term independent orbiting laboratory for other government agencies and perhaps orbital space tourism.

In common with the Space Shuttle, Dream Chaser will return to Earth by gliding, although, unlike the Shuttle, which was restricted to landing on two NASA-operated runways, Dream Chaser will land on any airport runway that handles commercial air traffic. The vehicle will feature a launch-escape system which will use its hybrid-motor pusher escape system to eject from a failing Atlas V, and fly a piloted return-to-landing-site maneuver with a 2-G load on the crew. For maneuvering in space, the vehicle will use its ethanol-fuelled reaction control system (RCS). The advantage of using ethanol, which is not an explosively volatile material, allows Dream Chaser to be handled immediately after landing, unlike the Space Shuttle, which always needed a small army to make it safe. Another improvement on the Shuttle is Dream Chaser's thermal protection system (TPS), an ablative tile created by NASA's Ames Research Center (ARC) that will be replaced as a large group rather than the laborious and time-intensive tile-by-tile process that was the case with the Shuttle. It is anticipated that parts of the TPS will only need to be replaced after several flights.

Publicly announced on September 20th, 2004, as a candidate for NASA's Vision for Space Exploration (VSE) and later for the agency's COTS, Dream Chaser was not selected under Phase 1 of the COTS program. This setback resulted in founder Jim Benson stepping down as Chairman of SpaceDev and starting Benson Space Company to pursue development of the vehicle. In April 2007, SpaceDev announced it had partnered with the ULA to pursue the possibility of utilizing Atlas V as the Dream Chaser's launch vehicle. This partnership was followed by SpaceDev's acquisition by SNC in December 2008. Then, on February 1st, 2010, SNC was awarded US$20 million in seed money under NASA's Commercial Crew Development (CCDev) 1 program for the development of Dream Chaser. Using this funding, SNC completed four planned milestones, including program implementation plans, manufacturing readiness capability, hybrid rocket test fires, and the preliminary structure design. Additional Dream Chaser tests included the drop test of a 15% scaled version (at the NASA Dryden Flight Research Center) to test flight stability and collect aerodynamic data for flight-control surfaces. Later that year, for the CCDev 2 solicitation by NASA in October 2010, SNC proposed extensions of Dream Chaser spaceplane technology. Following the CCDev 2 solicitation, SNC announced it had achieved two critical milestones for the program. The first consisted of three successful test firings of a single hybrid rocket motor in one day and the second was the completion of the primary tooling necessary to build the composite structure of the vehicle.

Six months after achieving these milestones, on April 18th, 2011, NASA awarded another US$80 million in CCDev funding to SNC. Following this funding, SNC completed nearly a dozen further milestones, including testing of the airfoil fin shape, integrated flight software and hardware, landing gear, and a full-scale captive carry flight test. These milestones were followed by a System Requirements Review, a new cockpit simulator, finalizing the tip fin airfoil design, and a Vehicle Avionics Integration Laboratory (VAIL), which will be used to test Dream Chaser computers and electronics in simulated space mission scenarios. Then, in June 2011, SNC signed a SAA with NASA, which was followed by ULA announcing that Atlas V would be used to launch Dream Chaser.

By February of 2012, SNC had completed the assembly and delivery of the primary structure of the first Dream Chaser flight test vehicle. With this achievement, SNC had completed all 11 of its CCDev milestones on time and on budget. Two months later, the company announced the successful completion of wind tunnel testing of a scale model of the Dream Chaser vehicle and, on May 29th, 2012, a captive carry test was conducted near the Rocky Mountain Metropolitan Airport, to determine its aerodynamic properties. This test was followed up by buffet tests on the Dream Chaser and Atlas V stack and aerodynamic and aerothermal analysis by the Langley/SNC team. On July 11th, 2012, SNC announced they had successfully completed testing of the Dream Chaser's nose landing gear during simulated approach and landing tests as well as the impact of future orbital flights.

Yet more funding followed on August 3rd, 2012, when NASA announced the award of US$212.5 million to SNC to continue work on Dream Chaser under the CCiCAP program. The 21-month contract began in August of 2012. With this funding, Dream Chaser is in a good position for achieving operational commercial human flight capability as early as 2016. Reaching this point had only been possible as a result of SNC having integrated the efforts of a powerful team of aerospace companies, academic institutions, and NASA centers to advance the development of Dream Chaser and the associated mission, ground, and crew systems. After completing a full-system PDR and the first captive carry flight, SNC now looks forward to more approach and landing tests, mirroring those of the first flight test of the Space Shuttle.

SNC has 10 CCiCAP milestones (Table 7.2) to meet during the 21-month base period. The company received what amounted to half an award that is significantly less than those provided to Boeing and SpaceX. The company also has 31 optional milestones for which it would receive funding if NASA has money available. These milestones were redacted from the SAA for competitive reasons.

Table **7.2.** SNC CCiCAP base milestones.

No.	Description	Date	Amount
1	The Program Implementation Plan Review is an initial meeting to describe the plan for implementing the CCiCap, to include management planning for achieving CDR; Design, Development, Testing, and Evaluation activities; risk management to include mitigation plans, and certification activities planned during the CCiCap Base Period	August 2012	US$30 million
2	The Integrated System Baseline Review (ISBR) demonstrates the maturity of the baseline CTS integrated vehicle and operations design of the Dream Chaser Space System consisting of the spacecraft, the Atlas launch vehicle, Mission Systems, and Ground Systems support	October 2012	US$45 million
3	The Integrated System Safety Analysis Review #1 demonstrates that the systems safety analysis of the Dream Chaser Space System has been advanced to a preliminary maturity level, incorporating changes resulting from the PDR	January 2013	US$20 million
4	The Engineering Test Article Flight Testing reduces risk due to aerodynamic uncertainties in the subsonic approach and landing phases and matures the Dream Chaser aerodynamic database; up to five Engineering Test Article free flight tests will be completed to characterize the aerodynamics and controllability of the Dream Chaser outer mold line (OML) configuration during the subsonic approach and landing phase	April 2013	US$15 million
5	The SNC Investment Financing #1 funding represents SNC's commitment for significant investing financing	July 2013	US$12.5 million
6	The Integrated System Safety Analysis Review #2 demonstrates the systems safety analysis of the Dream Chaser Space System (DCSS)	October 2013	US$20 million
7	The Certification Plan Review defines the top-level strategy for certification of the DCSS that meets the objectives for the ISS Design Reference Mission; SNC shall conduct a review of the verification and validation activities planned for the DCSS	November 2013	US$25 million

Table 7.2. *cont.*

No.	Description	Date	Amount
8	Wind Tunnel Testing reduces risk on the vehicle and the Dream Chaser/Atlas stack by maturing the DC and DC/Atlas aerodynamic databases, providing improved fidelity in Reynolds number effects and control surface interactions, and it will help determine pre-CDR required updates to the OML or control surface geometry if required	February 2014	US$20 million
9	The Risk Reduction and TRL Advancement Testing, whose purpose is to significantly mature all Dream Chaser systems to or beyond a CDR level	May 2014	US$17 million
9a	The Main Propulsion and RCS Risk Reduction and TRL Advancement Testing, whose purpose is to significantly mature the Dream Chaser Main Propulsion System and Reaction Control System to or beyond a CDR level; risk-reduction and Technology Readiness Level improvement tests will be completed for these systems	May 2014	US$8 million
Total:			*US$212.5 million*

In NASA's source-selection document, Dream Chaser was critiqued for posing significant risks due to design complexity. While the document acknowledged that a winged vehicle offered advantages to customers in terms of easier landings, lower Gs, and greater cross-range from orbit, NASA also pointed out that aborts and thermal protection issues were more difficult. It was due to these technology hurdles that the agency cut back on the milestones SNC had proposed, commensurate with the lower funding level. The good news is that SNC decided to go ahead with its bid to fly NASA astronauts with plans to conduct autonomous drop tests and landing with a full-scale engineering article. Ultimately, SNC believes, the capabilities of Dream Chaser will outweigh the risks stated in the source-selection document. The company's optimism is well placed; the hybrid motors confer a robust capability for servicing missions and the high reuse (25–30 missions each) that SNC expects are significant advantages over the Boeing and SpaceX capsules.

ATK

Profile
Spacecraft type:
Capsule with service module
Crew capacity: 7
Launch vehicle: Liberty, Atlas V or Delta IV Heavy
CCiCAP funding: None
Previous CCDev funding (including optional milestones): US$130.9 million
(Boeing), US$6.7 million (ULA)
Total CCDev and CCiCAP funding (if all milestones met): US$590.6 million
(Boeing), US$6.7 million (ULA)

Take a canceled, over-budget NASA rocket, stack a workhorse European satellite launcher on top of it, and the result could be an affordable alternative for ferrying astronauts into LEO. That's the proposal put forward by Alliant Techsystems (more commonly known as ATK), the aerospace company that used to manufacture the solid-rocket motors for the Space Shuttle, and Astrium, the European company that builds Ariane 5 rockets. ATK and Astrium's plan? To develop a commercial version of Ares I and rename it Liberty (Figure 7.8).

Ares I, a legacy of Constellation, a bloated NASA program to return astronauts to the Moon, could still become a success story in the Obama Administration's effort to develop a commercial space program. Liberty would be cheaper than Ares I, because the Ares I upper stage would be replaced by the Ariane 5 first stage which, at the time of writing, has been launched successfully 41 consecutive times. Liberty's lower stage, a longer version of the Shuttle booster built by ATK, would be almost unchanged from Ares I.

As one of the largest aerospace and defense companies in the US, with more than 18,000 employees, ATK has a long history in the manned and unmanned spaceflight arena. In addition to manufacturing the booster separation systems that released the motors away from the Shuttle orbiter and the main liquid fuel tank, nearly every NASA planetary probe has used some form of propulsion from one of the company's divisions. Like Orbital, Boeing, and SpaceX, ATK had hoped to receive funding through NASA's CCDev program but didn't make it beyond the second round. Undeterred, ATK continues development of the booster under an unfunded SSA with NASA. As part of this arrangement, NASA shares its expertise in designing and testing the rocket but does not provide money for the project.

The company's launch system is based on the Liberty rocket and a capsule that will carry passengers to destinations in LEO, such as the ISS and Bigelow habitats. Unlike NASA's Orion, which is largely built from aluminum lithium, the construction of ATK's crew module is based mostly on composites, one of the company's core competencies. While the look of the capsule will be similar to Orion, it will only perform flights to and from LEO, whereas Orion is designed for long-duration missions to asteroids and perhaps Mars. The capsule, which will land over water, will be reusable up to 10 times.

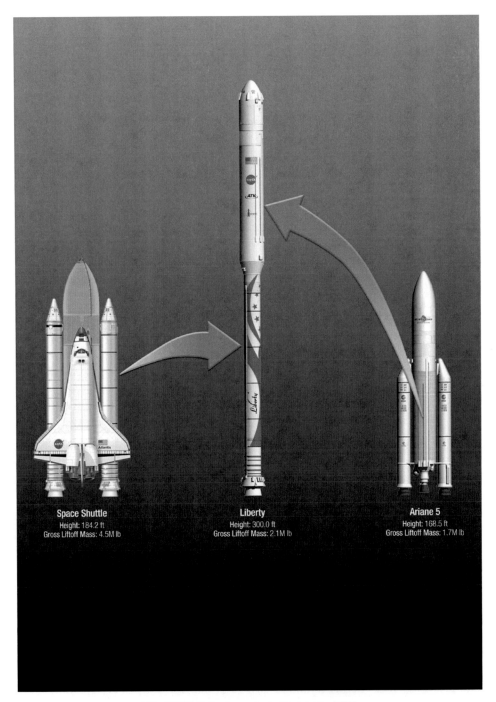

Space Shuttle
Height: 184.2 ft
Gross Liftoff Mass: 4.5M lb

Liberty
Height: 300.0 ft
Gross Liftoff Mass: 2.1M lb

Ariane 5
Height: 168.5 ft
Gross Liftoff Mass: 1.7M lb

7.8 ATK's Liberty rocket. Courtesy: ATK

ATK is planning to begin test flights of the Liberty system in 2014, with the first manned mission expected to occur in late 2015 – a schedule that may result in Liberty being available to NASA and other customers by 2016. ATK has already completed full-scale testing of Liberty's five-segment solid motor, which happens to be the world's largest solid-rocket motor and was originally designed to be the first stage of the Ares I rocket that NASA planned to use to launch the Orion capsule on trips to the Moon in the now scrapped Constellation Program. Liberty, which will tower 91 meters above the launch pad, will use the original Ares 1 engine as a first stage and the Ariane 5 rocket as the second stage. It will be capable of lifting more than any system currently available. Built in partnership with Astrium, whose Ariane 5 rocket has served as Europe's primary booster for launching satellites and spacecraft into orbit, and Lockheed Martin, who will provide support by designing avionics, navigation and control, and the vehicle's docking system, the launch system is advertised at a price of US$180 million per launch and has a projected payload of 20,140 kilograms to LEO.

While the ATK/Astrium proposal wasn't initially selected by NASA in the 2011 contract awards, the team has continued development in the hope of gaining funding from NASA in the future. To that end, ATK continues to work towards achieving the milestones required to develop the system. For example, in January 2012, ATK successfully held a Launch System Initial Systems Design (ISD) Review of the Liberty Transportation System (LTS), the third of five milestones to be completed under the SAA. Even without NASA funding, the LTS has a good chance of being successfully developed thanks to the partnership ATK has with Astrium. The collective launcher heritage of these two companies extends to nearly 150 successful flights and leverages billions of dollars of investments by NASA and NATO-allied European governments in the frame of the European Space Agency (ESA), but the advantages don't stop there. Take the five-segment first-stage design, for example. Based on more than 30 years of safety-driven improvements on the Space Shuttle program, the result is a high-performing, very reliable solid-rocket motor, which equates to increased safety for crew and mission success for cargo. Besides adding a fifth segment, ATK also enhanced the propellant grain, provided a larger nozzle opening, and upgraded the liner and insulation – all designed to meet performance requirements and increase reliability while significantly lowering manufacturing costs.

ATK's development timeline stretches back to September 2009, when the company ground-tested the five-segment first stage. The following month demonstrated the vehicle's proof of concept following the successful Ares I-X flight test – a launch vehicle configuration very comparable to Liberty. The Ares I-X flight also demonstrated effective vehicle integration, ground-processing, and launch operations.

On paper, ATK had a strong team helping them. For example, EADS is a global aerospace and defense giant; in 2009 alone, the group – comprising Airbus, Astrium, Cassidian, and Eurocopter – generated revenues of €42.8 billion and employed a workforce of more than 119,000. Snecma (Safran Group), meanwhile, is one of the world's leading manufacturers of aircraft and space engines, and offers a wide range of propulsion systems. Snecma also develops and produces propulsion systems and equipment for launch vehicles and satellites and brings together 40 partners across 12 of ESA's member states, to produce Ariane's Vulcain, Vulcain 2, and HM7B engines. The Ariane 5 launcher is developed and manufactured by Astrium and operated by Arianespace, which is the world's leading launch company, backed by 21 shareholders and ESA.

ATK had cause to be optimistic at the start of 2012 after it completed its third SAA Milestone (Launch System ISD Review) for Liberty under NASA's CCP. The SAA enabled NASA and ATK to share technical information related to the LTS during the PDR phase of the program. During this meeting, ATK presented the status of Liberty's system-level requirements, preliminary design, and certification process. All efforts to date had been supported exclusively by internal funding but the ATK team was optimistic, since their vehicle had the highest pounds-to-orbit of any other vehicle currently working under commercial agreements and it was the only vehicle originally designed for human rating.

That optimism was still evident four months later when ATK announced it had developed Liberty into a complete commercial crew-transportation system, including the spacecraft, abort system, launch vehicle, and ground and mission operations, designed from inception to meet NASA's human-rating requirements with a potential for the first test flight in 2014 and Liberty crewed flight in 2015. The company also announced that Lockheed Martin would provide support to the ATK and Astrium Liberty team as a major subcontractor on the project. Under the agreement, Lockheed Martin would provide crew interface systems design, subsystem selection, assembly, integration, and mission operations support for the Liberty vehicle.

In June 2012, development was still on track, as ATK announced it had completed its Liberty software technical interface meeting (TIM), which was held to support further development of the Liberty space transportation system under its SAA. The software TIM evaluated Liberty's software development plan with the NASA Liberty team. The plan, which governs the software process used by Liberty and its subcontractors throughout development, integration, test, and flight, is critical for understanding the entire system to support Liberty's test flights.

The following month, the Liberty program announced an independent assessment team led by ex-astronaut Bryan O'Connor. O'Connor's first task was to advise ATK on development of its commercial human certification plan for the Liberty system, which includes the launch vehicle, upper stage, abort system, composite spacecraft, ground and mission operations, crew and passenger training, and a test flight crew. Helping O'Connor with man-rating the Liberty system will be Ken Bowersox, a former navy test pilot and Shuttle astronaut with four Shuttle missions and one ISS mission. During his tenure at NASA, the former ISS Commander held a variety of assignments, notably chief of the Astronaut Office Safety Branch and chairman of the Spaceflight Safety Panel.

The same month, ATK announced an extended cargo configuration allowing the Liberty spacecraft to transport a pressurized pod (the Liberty Logistics Module or LLM) together with the composite crew module. Based on NASA's Multi-Purpose

Logistic Module design, the LLM will include a common berthing mechanism and will be capable of transporting up to 5,100 pounds of pressurized cargo, equivalent to carrying four full-sized science racks, to the ISS.

ATK's final SAA Liberty milestone – a Program Status Review (PSR) – was completed late July 2012. During the PSR, the Liberty team presented NASA with detailed progress of the program, including an integrated master schedule, system requirements, software status, flight test plan, system safety review, ground-processing certification plan, and schedule for initial operation capability. All seemed to be progressing well, until NASA announced the company had been overlooked for CCiCap funding.

It was a serious blow to ATK, who had invested significantly in its push for a CCiCap award. Given this investment, it was hardly surprising the company felt somewhat optimistic about its chances of winning CCiCap funding – optimism that no doubt prompted ATK's Kent Rominger to outline an aggressive schedule for Liberty that called for flight tests in 2014 and crew flights starting in 2015. The schedule, announced at the 2012 NewSpace conference, depended on winning a full CCiCap award so, when the CCiCap funding was announced, ATK had to recalibrate.

Some space observers argued that unless ATK could find customers for their rocket, it would be difficult for them to stay in business long enough to do anything with their spacecraft. At the time of the CCiCap announcement, no market niche existed that the Liberty rocket addressed better than other existing operational rockets. And if/when SpaceX's Falcon Heavy becomes operational, it will offer more capacity for about two-thirds of the price. How will ATK gain a market share against that sort of competition? While many didn't see a bright future for the Liberty rocket even if they had won a CCiCap award, ATK indicated it would continue, but at a slower pace, and planned to re-evaluate its program, which might result in developing the Liberty rocket as a launcher for satellites, while holding open the possibility of launching crews at a later date.

ORBITAL

Profile
Spacecraft: Cygnus – Cargo Freighter
Type: Capsule with service module
Launch vehicle: Antares two-stage
Previous CCDev funding (including optional milestones): US$130.9 million (Boeing), US$6.7 million (ULA)
Total CCDev and CCiCAP funding (if all milestones met): US$590.6 million (Boeing), US$6.7 million (ULA)

Orbital is an American company specializing in the manufacture and launch of satellites. The company, which has built more than 500 launch vehicles and almost 200 satellites, has a 40% share of the interceptor market, 55% share of the small communications satellite market, and a 60% share of the small launch vehicles market. Founded in 1982 by David Thompson, Bruce Ferguson, and Scott Webster, the company first broke into the commercial space industry in 1990, when it successfully carried out eight space missions, highlighted by the launch of the Pegasus rocket.

Until recently, Orbital's space launch vehicles have focused on boosting small payloads to orbit; its workhorse, the Pegasus rocket, which is launched from the company's L-1011 carrier aircraft, *Stargazer*, has conducted 40 missions. The company also provides suborbital launch vehicles for the nation's missile defense systems. In short, Orbital is a versatile company, so it's not surprising it has designs on breaking into the commercial manned spaceflight market, which is the goal of its Advanced Programs Group. In support of pursuing its human spaceflight goals, Orbital is developing Antares, a medium-lift rocket, and Cygnus, a spacecraft that will provide supplies to the ISS. Both programs are being developed under NASA's COTS, for which Orbital was selected as a partner in 2008. Scheduled to begin operational flights by the end of 2012,[1] the company will provide resupply services to the ISS under NASA's Commercial Resupply Services (CRS) program from the new Mid-Atlantic Regional Spaceport at Wallops Island in Virginia.

In 2010, in response to NASA's CCDev 2 solicitation, Orbital made a commercial proposal to develop a lifting-body VTHL vehicle. The VTHL, about one-quarter the size of the Space Shuttle, would be launched on a human-rated Atlas V rocket and would seat a crew of four. After failing to be selected for the CCDev 2, Orbital announced in April 2011 that they would likely wind down their efforts to develop a commercial crew vehicle.

The COTS collaboration between Orbital and NASA will involve full-scale development and a flight demonstration of a commercial cargo delivery system scheduled to take place in 2013. The COTS system consists of Antares, Orbital's launch vehicle, Cygnus, the company's spacecraft, and pressurized modules to transport cargo to the ISS. Under the CRS program which awarded Orbital Sciences a US$1.9 billion contract on December 23rd, 2008, Orbital will deliver up to 20 tonnes of cargo to the ISS through 2016 in eight Cygnus flights, utilizing its Antares, Cygnus, and a Pressurized Cargo Module (PCM) developed by Orbital's industrial partner, Thales Alenia Space.

Orbital will design, manufacture, and test its Antares launch vehicle in Dulles, Virginia, and Chandler, Arizona, while the development, production, and integration of the Cygnus spacecraft and cargo modules will be done in Dulles and the Wallops Flight Facility, where COTS/CRS launches are planned. The

---

[1]   Given Orbital's recent run of launch failures, NASA will no doubt be hoping for a change in the company's fortunes; on March 4th, 2011, the agency's Glory atmospheric research satellite was lost due to a mechanical failure on an Orbital Taurus XL rocket. This failure followed the February 24th, 2009, failure of the Orbiting Carbon Observatory due to a similar Taurus XL malfunction. The estimated loss to NASA for the failures was close to US$700 million.

Antares launch vehicle will have a payload capacity of 5,000 kilograms to LEO, while the Cygnus spacecraft will be capable of delivering up to 2,700 kilograms of pressurized cargo to the ISS.

The Cygnus spacecraft (Figure 7.9) comprises two components: the PCM and the Service Module (SM). The first iteration of the PCM, which is manufactured by Thales Alenia Space, will have a volume of 18 cubic meters, while the SM is built by Orbital and is expected to have a gross mass of 1,800 kilograms with propulsion provided by thrusters using the hypergolic propellants hydrazine and nitrogen tetroxide. The enhanced variant of the Cygnus, which will use a stretched PCM, increasing the interior volume to 27 cubic meters, will be mated to a new upper stage, the Castor 30XL, which will boost the payload Cygnus can deliver to the ISS by 700 kilograms.

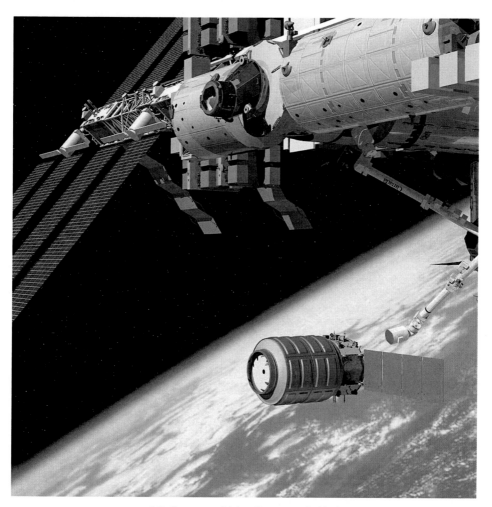

7.9 Cygnus vehicle. Courtesy: Orbital

In common with SpaceX's Dragon cargo missions, the Cygnus CRS mission will maneuver close to the ISS, where the Canadarm2 robotic arm will grapple the vehicle and berth it to a Common Berthing Mechanism on the Harmony module. Nominally, the Cygnus will remain berthed for about 30 days. Unlike Dragon, the Cygnus doesn't provide return capability, although it can be loaded with obsolete equipment and trash for destructive re-entry similar to the Russian Progress vehicles.

In addition to the Cygnus, Orbital is also working on readying their Taurus 2 rocket (the booster for Cygnus flights), preparing the unmanned Cygnus cargo freighter for its demonstration flight and readying the construction and certification at a new NASA launch site. Of these three development tracks, the pacing item is the launch site; once this is certified, Orbital will be ready to move into full operation.

In July 2009, Orbital announced the award of a contract to Thales/Alenia Space of Turin to build nine PCMs that will deliver supplies to the ISS. This agreement followed a contract signed with Applied Space Structures of Stockton, California, to provide composite structures for the Cygnus SMs. The following month, the company conducted its CDR of the Cygnus PCM, a joint effort between Orbital and Thales Alenia Space Italia. The CDR demonstrated that the PCM design was ready for full-scale fabrication, assembly, integration, and test of the modules. After conducting the CDR, Orbital presented their CCDev 2 documentation to Johnson Space Center's (JSC) Safety Review Panel to ensure all appropriate safety requirements had been met to ferry astronauts. This was the first in a three-part process to ratify the Cygnus for human spaceflight. At the close of 2009, Orbital signed a contract with Dutch Aerospace to supply the solar arrays that will power the Cygnus. In April 2010, Orbital displayed a full-scale model of the Cygnus cargo vehicle at the National Space Symposium in Colorado Springs while the first PCM was taking shape at the Thales/Alenia facility. Later that year, Orbital added a new flight to their manifest – a risk-reduction test flight that would carry a payload simulator in place of the full-fidelity Cygnus spacecraft to verify the design and flight performance of Taurus II. In August, the PCM successfully completed pressure testing – a test that was followed by the NASA/Orbital Joint Avionics Test at the NASA Station Development and Integration Laboratory near JSC. The purpose of the test was to ensure the Cygnus and ISS flight software were capable of communicating via berthed and proximity communications system links. The test also demonstrated basic command and telemetry data routing between the ISS and Cygnus flight software. The test demonstrated 16 of the 21 design-verification objectives required to satisfy NASA requirements. In October, the Cygnus SM successfully completed its static load test at Applied Aerospace Structures. The test applied loads to the structure to simulate the forces the vehicle will experience during a Taurus II launch while carrying a fully loaded PCM on top. In November 2010, NASA Administrator Charles Bolden and Orbital's CEO David Thompson performed a ribbon-cutting ceremony to mark the completion of Orbital's Mission Operations Center for the company's cargo missions. At the end of 2010, Orbital and Thales Alenia Space successfully completed a COTS program milestone by performing a Cargo Integration Demonstration, which involved volumetric loading of a standard PCM to demonstrate access to all the cargo stowage locations. Its next COTS milestone was achieved in June 2011, when the PCM completed Qualification and Hardware Acceptance Reviews at the Thales/Alenia facility in Turin. This cleared the way for the PCM to be transported to Wallops Flight Facility in Virginia. In August 2011, the Cygnus PCM was ferried aboard an Antonov An-26 aircraft to Wallops before being mated to the Cygnus SM at NASA's H-100 payload-processing facility. The PCM and SM, which will form the first operational Cygnus, were now ready for more tests, including static fire tests, environmental testing, and thermal vacuum testing, which continued into 2012.

As this book is being written, Orbital was preparing for a February 2013 test flight of the Antares with a Cygnus mass simulator. Unlike Dragon's October 2012 flight to the ISS, this flight won't rendezvous and berth with the ISS. However, unlike SpaceX, which had limited experience building spacecraft, Orbital's near-30-year history of building commercial satellites should help it speed up these efforts.

BLUE ORIGIN

Profile
Spacecraft: New Shepard (suborbital), Space Vehicle (orbital)
Type: Capsule with service module
Crew capacity: 7
Launch vehicle: Atlas V (Blue Origin has plans to build its own reusable launch vehicle)
CCiCAP funding: None
Previous CCDev funding: US$25.7 million in the first two rounds

Founded in 2000, Blue Origin is a private company that has a long-term vision of significantly increasing the number of people who fly into space through low-cost and reliable commercial space transportation. Little is known about what goes on at the Seattle-area company, with news only leaking out a few times a year, so details about the innovative design the company is developing for its creatively (!) named Space Vehicle (SV) are sketchy. But why is the company so tight-lipped about its work? Well, first of all, Blue Origin likes to talk about things they have done and not things they are planning to do. In other words, Blue Origin is a company focused on accomplishments. In a business as hard as the space business, in which things usually take (much) longer than expected, such a philosophy is difficult to argue against. Another reason for Blue Origin's secret squirrel approach is that the company doesn't want to lose focus. Again, such an approach makes sense; after all, the more time spent talking, the less time there is to spend doing.

Blue Origin's CTS comprises the aforementioned SV (Figure 7.10), which, to begin with, will be launched on an Atlas V launch vehicle, but only until Blue Origin's own Reusable Booster System (RBS) is developed. The biconic SV will be capable of carrying seven astronauts and, just like Dragon, it will transfer NASA crew and cargo to and from the ISS, serve as an ISS emergency escape vehicle for up

7.10 Blue Origin's New Shepard vehicle. Courtesy: Blue Origin

to 210 days, and perform a land landing. It will also be available to conduct separate commercial missions for science research and private adventure, and be capable of traveling to other LEO destinations. While the SV is designed to ride on multiple boosters, Atlas V was chosen because it has a proven launch track record, has the required performance capability, can be adapted for human spaceflight operations, and is operated from facilities close to the Kennedy Space Center.

As previously mentioned, Atlas V is a temporary measure, since Blue Origin is simultaneously developing its RBS. The rationale behind the RBS is simply stated; over the next few years, NASA will use a lot of expendable booster stages and one-time-use expendable booster technology is very expensive. A much cheaper alternative is Blue Origin's RBS, which employs deep-throttling, restartable engines to perform VTVL maneuvers for booster recovery and reuse.

Thanks to its US$22,005,000 in CCDev 2 funding, Blue Origin is pursuing three projects, one of which is maturing the SV design. The company is steadily advancing its SV design through completion of key system trades such as the design of the thermal protection system, defining the biconic shape including aerodynamic analyses, CFD analysis, and wind tunnel testing, developing a draft SV, and completing a Mission Concept Review (MCR) and System Requirements Review (SRR), resulting in a baseline definition architecture and system requirements.

The second of the company's projects focuses on maturing the SV's Pusher Escape System, which will provide the vehicle's crew escape capability in the event of an anomaly during launch. The pusher escape system, which uses an engine in a pusher configuration, is a technical risk in Blue Origin's SV concept, since it differs markedly from the traditional towed-tractor escape tower concepts utilized on

Mercury, Apollo, and Orion. One of the key technologies required to develop the pusher escape configuration is active thrust vector control (TVC), which will be used to steer the SV away from the lower stage during ascent. Blue Origin has already conducted TVC ground testing, which was completed under CCDev 1. Additional testing will include a pad-escape abort test and perhaps a high-dynamic-pressure (Max Q) abort test.

The third project Blue Origin is working on is accelerating development of its 100,000 pounds of thrust liquid-oxygen/liquid-hydrogen (LOX/LH2) engine. Blue Origin's restartable, deep-throttle engine is the pacing item for the company's orbital RBS program. Under CCDev 2, the company proposed testing the full-scale thrust chamber at NASA's Stennis Space Center and conducting development testing of the engine's fuel and oxidizer turbo-pumps.

In an effort to better understand and characterize their system, Blue Origin has adopted an incremental development approach that uses a reusable suborbital vehicle (New Shepard) capable of flying three or more astronauts to an altitude of above 100 kilometers for science research, adventure, and revenue. The suborbital capsule, which will separate from the subscale booster prior to re-entry, followed by a land landing for recovery and reuse, will baseline key technologies for the orbital SV and, as this book is being written, is undergoing final assembly.

In common with all the companies engaged in the space taxi race, Blue Origin has assembled an experienced team, including NASA's ARC, for system trades, design, and test activities related to SV design maturation, Stennis Space Center (SSC), for engine thrust chamber testing, and ULA, for integration of the SV with Atlas V. Other team elements include major suppliers such as Aerojet, for solid-rocket motors and test facilities, Lockheed Martin Missiles & Fire Control High Speed Wind Tunnel (HSWT), for high-speed wind tunnel SV testing, and US Air Force Holloman High Speed Test Track (HHSTT), for pusher escape system testing.

WINNING THE SPACE TAXI RACE

Who will win the space taxi race? It's difficult to say but it will probably be one of the companies awarded CCiCap funding. The others? ATK may have an uphill struggle, according to a NASA source-selection document released on September 4th, 2012, which indicated the company's design was dropped from NASA's shortlist of potential space station taxis because the company didn't present a technically sound plan for combining existing rocket and spacecraft designs into a single transportation system. Another possible strike against ATK is the desire to field a strictly American launch vehicle and, while Liberty does employ an American first stage, albeit based on a solid-rocket booster, the second stage is entirely European.

As it stands in mid-2013, SpaceX's Falcon 9–Dragon combination represents the only all-American launch solution on the table, designed and built in California, tested in Texas, and launched in Florida. For a US workforce desperate for jobs, and a country looking for something to celebrate, it doesn't get much better than that. The company also happens to be at the forefront of the space taxi race, investing its own resources to build what everyone agrees is necessary to establishing a permanent future in space: a fully reusable space transportation system. If all goes well, that system could be in service by 2017. All might not go well, of course. Delays are possible, even probable, especially in this industry, as SpaceX knows only too well. But that date may be important because NASA's Orion multi-purpose crew vehicle is on a go-slow development path to free funds for near-term agency objectives. While not part of the commercial space taxi race, Orion was initially intended to serve as transportation to the ISS and later for missions beyond LEO. Now, with the commercialization of transport to the station as the preferred route, Lockheed Martin – Orion's prime contractor – is focusing more on deep-space exploration, although the vehicle will still be able to perform the station mission. Slated for an unmanned mission atop the Space Launch System (SLS) in 2017, Orion will perform three test flights, the third of which will be a manned flight planned for 2021. While Orion may not serve as an ISS ferry, the program has benefitted from changes in NASA oversight that has grown out of the commercial crew programs, which are conducted under Space Act procedures that are less restrictive than previous agency procurements. For example, Lockheed has NASA engineers working side by side with its engineers in Orion's design and testing work – a move that gives the agency insight into Lockheed's work progress. The arrangement has advantages for Lockheed as well, since it benefits from NASA's experience in water-landing tests and arc-jet procedures.

On the flip side, the aerospace supply base is shrinking, which is having an impact on low-volume procurements for Orion's systems, especially in the supply chain for radiation-hardened electronics. One way of accelerating Orion's development would be to work with ESA to develop a SM based on the Automated Transfer Vehicle (ATV), but this proposal has run into opposition from some ESA partners, which means that, if the US wants to launch astronauts between now and 2017, it will have to continue relying on the Russians or hope that Dragon is man-rated ahead of time. That should concern people. No doubt it does. But, with the Shuttles retired without a replacement, the US simply doesn't have another choice.

8

Red Dragon

"Ultimately, the thing that is super important in the grand scale of history is 'Are we on the path to becoming a multi-planet species or not?' And if we're not, that's not a very bright future, we'll simply be hanging out on Earth until some eventual calamity claims us."

Elon Musk at a conference held on August 2nd, 2011,
by the American Institute of Aeronautics and Astronautics

As Curiosity, the US$2.5 billion Mini Cooper-sized Mars rover, touched down on the Red Planet on August 3rd, 2012, Elon Musk was already planning the next logical step – sending humans there. And, as with all of Musk's space plans, his goal wasn't short on ambition; he wasn't just interested in ferrying people to Mars, but making it possible for people to live there ... permanently. Musk acknowledges that one of the biggest challenges of colonizing the Red Planet is making the trip affordable for the average American, suggesting a price point of a round-trip ticket should be no more than half a million dollars. It's a bold plan, especially when you consider most manned Mars missions are budgeted in the multi-billion-dollar range. But "bold" fits well with the man who created PayPal, Tesla, and SpaceX.

Manned missions to Mars have always been a popular spaceflight topic, and not just in the wake of Curiosity's arrival. In September 2011, NASA unveiled the Space Launch System (SLS) that the Obama Administration hopes will deliver humans to the Red Planet by the 2030s. Then there's the Mars One project, which is planning to send their astronauts on one-way trips to Mars starting in the 2020s – possibly using SpaceX hardware. But, despite all the discussion about sending humans to the Red Planet, there have been few people more vocal than Musk, who believes his company can land humans on Mars in 12–15 years. If he's right, it's entirely possible that Curiosity could still be roving around when the first astronauts land.

If getting humans to Mars in 12–15 years sounds ambitious, that's because it is. In the weeks leading up to Curiosity's arrival on the Red Planet, the public was inundated with press releases from NASA explaining how difficult it is to get a robot there; getting a human there is a completely different ballgame. For one thing, humans require bulky life-support systems, food, and living space. Musk understands the myriad inherent dangers but, while he acknowledges the first flight will be risky, he also has no qualms about being on the first flight as long as he feels comfortable that SpaceX is in the right hands ... just in case.

If it was anyone else talking about how to send people to Mars for half a billion dollars each, then one would have to be skeptical but, when it's Musk talking, it's difficult to be anything but guardedly optimistic, especially when you consider SpaceX's track record of getting things done. The company already has the vote of confidence from NASA, which selected SpaceX to return American astronauts to space, awarding the company US$440 million CCiCap funding. SpaceX's fleet of space taxis should launch in 2017. The question is how soon can SpaceX follow it up with a crewed mission to Mars?

The key to achieving Musk's interplanetary dream lies in SpaceX's principle of eliminating equipment costs for space travel. By realizing total reusability and leaving fuel as the only financial burden, SpaceX hopes to make spaceflight more affordable. It's a principle that can be applied to a manned mission to Mars.

One of the keys to a SpaceX Mars mission will be the Falcon 9 Heavy (Figure 8.1). Capable of putting more than 53 tonnes of payload (more than twice that of the Space Shuttle) in low Earth orbit (LEO), the Falcon Heavy will be transformative because it will substantially reduce the cost of carrying a given mass

Launch Vehicle Evolution

- Falcon 1 → Falcon 1e (2011)
  - 1050 kg to LEO
- Falcon 9/Dragon crew transportation (~30mths after ATP)
- Falcon 9 → Falcon 9 Heavy (net 2013)
  - 32k kg to LEO
- Merlin 2 booster engine:
  - ~1.7M lbf LOX/RP-1
- Raptor upper stage engine
  - LOX/LH2
- Falcon X
  - All RP Heavy Lift
  - 38k kg to LEO
- Falcon X Heavy
  - All RP Super Heavy Lift
  - 125k kg to LEO

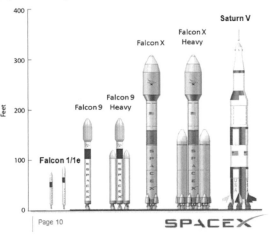

8.1 SpaceX launch vehicle evolution, showing the Falcon Heavy compared with other launch vehicles. Courtesy: SpaceX

into space – perhaps as low as, or lower than, US$1,000 per pound lifted. Other important contributions to the Falcon Heavy's low cost are the advanced design of its structure, its avionics, its engines, and its launch operation (Appendix IV). But, while these innovations are important, what will ultimately slash space access costs will be the vehicle's reusability.

One of SpaceX's first steps towards a manned mission to Mars is an unmanned sample return mission that will test the techniques and technologies that will be required in a manned mission. The sample return mission proposed by SpaceX will use a Dragon variant – the appropriately named Red Dragon – to ferry instruments such as a drill that will bore about one meter underground to sample reservoirs of water ice (potential landing sites would be polar or mid-latitude sites with proven near-surface ice known to exist below the surface). NASA's Ames Research Center (ARC) is working with SpaceX to plan a mission projected to cost less than USD$400 million, plus about US$130 million for one Falcon Heavy. In addition to the search for life, assessing subsurface habitability, and establishing the distribution of ground ice, the Red Dragon mission will also conduct a human-relevant entry, descent, and landing (EDL) test, demonstrate access to subsurface resources, and conduct in-situ resource utilization (ISRU) tests – all important technologies that need to be tested before a manned mission gets off the ground.

While the Red Dragon mission, if approved, will help SpaceX's mission planners better define the technologies required by a manned mission, the EDL challenge will remain. Thanks to its design, Dragon can probably perform the necessary EDL functions to deliver a payload of one tonne to the Martian surface without using a parachute (using parachutes is not feasible without vehicle modifications that would require a significant development program) because the vehicle's drag may slow the capsule sufficiently for the descent to be within retro-propulsion capabilities. Apart from avoiding the use of risky hover-drop techniques, such an approach will probably make it possible to land the capsule on legs at much higher Martian elevations than could be done if a parachute was used. Such an approach requires the use of retro-propulsion, which is why SpaceX is developing the SuperDraco engine.

Integral to SpaceX's future unmanned and manned interplanetary aspirations, the SuperDraco is a powerful new engine currently under development that will enable the capsule to land on solid ground rather than dropping into the ocean. For Dragon's May 2012 flight to the International Space Station (ISS), Dragon was powered by SpaceX's previous-generation engine, the Draco. The Draco thrusters generate 90 pounds of thrust and Dragon used 18 of them to maneuver around the ISS while performing its approach and rendezvous exercises. The SuperDracos are a different animal altogether, generating an impressive 15,000 pounds of thrust (by comparison, the Apollo Lunar Module's ascent engine created 3,500 pounds of thrust). With a cluster of eight, this isn't enough power to take off from Earth, but it is enough to allow Dragon to land on Mars, take off again, and then return to Earth and perform a powered landing.

The SuperDraco, a prototype of which has been tested at SpaceX's Rocket

Development Facility in McGregor, Texas, is unique because it uses modified kerosene rather than standard solid fuel. Not only is this cheaper, it also allows the engines to be shut off by astronauts in the event of an emergency, allowing a launch to be aborted at any time, unlike the Space Shuttle's solid boosters which, once lit, couldn't be switched off. It's a design that also means the engines can be restarted multiple times and used repeatedly. Another advantage is that the engines won't have to be completely stripped and re-serviced after every launch. Here's how the engines might be used to decelerate a vehicle – a Red Dragon presumably – in a potential mission to Mars. The SuperDraco engines would decelerate Red Dragon from supersonic speed to a soft landing on the Martian surface. As it reached the Martian atmosphere, Red Dragon would be traveling at six kilometers per second and would utilize entry guidance during the hypersonic and supersonic phases of flight. Rather than deploying a parachute decelerator, Red Dragon would transition directly from atmospheric flight to powered descent at Mach 2.24. A high-thrust, powered descent phase would rapidly slow the vehicle to 2.4 meters per second about 40 meters above the surface. From that height, the vehicle would descend at 2.4 meters per second, before landing on a legged subsystem.

SpaceX and NASA have already studied such an EDL sequence in some detail. Based on a 7,200-kilogram vehicle,[1] the (unmanned in this case) Red Dragon would be configured by a center-of-gravity (CG) offset to fly with a lift/drag (L/D) of 0.24. Just before the powered descent phase, 120 kilograms of ballast mass (a potential science package mission of opportunity) would be ejected to remove the CG offset, thereby balancing the vehicle. During the powered descent, more mass would be shed due to the 1,900 kilograms of propellant that would be required to provide the power (or delta V) for a soft landing. The system total landed mass therefore would be 5,180 kilograms. Now, this is far short of the anticipated 30-metric-tonne payload mass of a manned mission, but the Red Dragon mission would be helpful in promoting into implementation the key technology of supersonic retro-propulsion.

While SpaceX is still some way from building a spacecraft that can take off and navigate its way back to a launch pad under its own power, the SuperDraco is a step in the right direction to dramatically lowering the cost of space access to the point at which a mission to the Red Planet may very soon become a realistic technological and financial prospect.

Speaking of finances, Musk is a business man and a very successful one, so it stands to reason he will want to justify a trip to Mars from a business standpoint. After all, given he will be making a huge investment, he will probably want a big return somewhere down the line. How? First, let's consider bragging rights. How much money would someone be willing to pay to be able to say they were the first to set foot on the Red Planet? That's the kind of immortality that would be incalculable. Or, how much would wealthy vacationers be willing to spend for the

---

[1] This is at or above the capability of the Delta IV Heavy, which is the largest launch vehicle in NASA's stable; SpaceX's Falcon Heavy, when ready, will be able to launch 7,200 kilograms to Mars with room to spare.

ultimate high-end destination? US$100 million? US$200 million? If those numbers sound excessive, consider the fact that Space Adventures is selling a round trip to the Moon (no landing, incidentally) for a cool US$150 million. A trip to Mars for a similar price would be a bargain of a lifetime.

Another potential return on investment might lie in mining. It's highly likely that Mars will turn out to have valuable minerals that are worth going there to mine, although, to make good on the investment, you would have to spend the money to ship the stuff back to Earth. Another potential cash cow is the discovery of life. There's good reason to suspect Martian life will actually resemble Earth life, which could prompt new lines of bio research. Then there's the possibility of patenting the genes of a non-Earth species, which would surely be worth big bucks.

Another potential source of revenue is reality TV, which brings us to the subject of the aforementioned Mars One, an eyebrow-raising and extremely ambitious (some say too ambitious) interplanetary project that plans to populate the Red Planet with reality TV show contestants. Yes, you read that correctly; the Dutch company Mars One plans to select several teams of four astronauts and the public will judge which team will get the ticket for a (one-way!) seven-month trip to the Red Planet in around the 2023 timeframe. The reason for mentioning it in this book is because Mars One intends to use SpaceX hardware to ferry their contestants to their interplanetary destination, although it should be noted that there is no formal agreement in place between Mars One and SpaceX. In fact, it is because Mars One intends to use SpaceX technology (and other existing hardware) that some think the venture has a fighting chance. Certainly, simplicity seems to be the driving force behind the Mars One proposal, since choosing suppliers on the basis of price and quality rather than political or national preferences, like other agencies may do, should save money. Another cost-saving measure is the reliance on tried and trusted technologies rather than the development of new ones. By sidestepping a lot of expensive research and testing, Mars One reckons they can keep the cost of the mission down to about US$6 billion to settle the first four astronauts; in contrast, the total cost of the Mars Science Laboratory that landed Curiosity was about US$2.5 billion.

Still, it's not surprising that many doubt the feasibility of such a project, not just because of the tight timeframe, but also because the bills of the Mars One suppliers will come before the company has had an opportunity to generate significant income. And how can a reality show expect to raise US$6 billion dollars? Well, Mars One hopes to spin the mission into the biggest media event ever – an event that will focus particular attention on the selection of the astronauts and their training. The company reckons this tactic will convince sponsors and investors to participate with the promise of more exposure later. We'll see.

How will the whole enterprise work? First there will be an astronaut-selection process (Table 8.1), and those selected will be placed in groups of four and subjected to a 10-year training program that will familiarize them with equipment and mission procedures. After a decade of training, the contestant astronauts will blast off on top of a Falcon Heavy, ensconced in a Red Dragon, where they will spend the next five or six months. Once they arrive at the Red Planet, all consumables will be waiting for them thanks to eight unmanned missions[2] that will have delivered food, water, and supplies in advance of their arrival. Every two years, another group of four contestants will join the settlement, expanding capabilities, bringing new supplies, and creating more reality-show-worthy interpersonal dynamics amongst the crew. This incremental expansion of the settlement will continue until the population becomes 40 strong. Then, if everything is still going according to plan (and presuming nobody has strangled one of their fellow contestants), a new village might be created at a different location on Mars. Eventually, in the distant future, the local population might reach the numbers necessary to build its own rocket and allow some of the astronauts to come back to Earth. Like I said, it's an ambitious plan.

Table 8.1. Mars One roadmap.

Year	Event
2013	‣ Astronaut selection begins
	‣ Mars One builds replica of the settlement on an Earth desert to help the astronauts prepare and train in a realistic environment
	‣ Astronaut selection and preparations in the simulated Mars base broadcast on television for the public to view and select the final four
2014	‣ Preparations for the supplies mission and for the first Mars communications satellite
2016	‣ Supply mission using a modified SpaceX Dragon launched to Mars in January 2016
	‣ Supply mission lands on Red Planet in October 2016 with its cargo of 2,500 kilograms of food
2018	‣ Robotic exploration vehicle lands on Mars to join the supply lander
	‣ Rover travels around Mars to determine optimum location for settlement
2021	‣ Settlement components comprising modified Dragon capsules reach their destination: two living units, two life-support units, a second supplies unit, and a second rover
	‣ Rovers take all components to settlement location and prepare for arrival of astronauts
2022	‣ Water, oxygen, and atmosphere production ready
	‣ Earth crew gets a go-ahead for launch
	‣ Components of the Mars Transit Vehicle launched into a low orbit, and linked together
	‣ On September 14th, 2022, the first four astronauts launched on top of a Falcon Heavy
2023	‣ Astronauts land on Mars
	‣ Astronauts link landers together
	‣ Astronauts set up remaining solar panels and begin exploration
2025	‣ Second group lands
2033	‣ Colony reaches 20 settlers

---

[2]   This may be reassuring to the contestants, since the landing sequence will already have been successfully performed eight times by the identical, unmanned capsules.

8.2 Mars One concept of operations. Courtesy: Mars One

Mars One has developed a technical plan that seems simple enough to be achieved within the next decade thanks to the hardware being readily available or at least in development. Unlike SpaceX, Mars One is not an aerospace company and won't produce any space hardware. Instead, the company will order their equipment from suppliers like SpaceX. In the Mars One scheme of things, Red Dragon will be utilized as a supply unit, a living unit, a life-support unit, a lander for the astronauts, and a lander for the rover. The first four of these variants will be linked using a passageway (Figure 8.2).

The supply unit will, as its name suggests, contain supplies, while the life-support unit will provide the astronauts with energy, generated by thin-film solar panels, and water and oxygen, which will be extracted from water ice in the Martian soil. The Mars One planners estimate the life-support unit will collect about 1,500 liters of water and 120 kilograms of oxygen in 500 days. The lander, which hosts the life-support unit, also serves as the living unit – one that has an inflatable living section and an airlock, which will be used by the astronauts when leaving the settlement. The lander also contains construction materials and the "wet areas", such as the shower and kitchen. In contrast to the scientific rovers such as Curiosity, the Mars One Rover will focus more on utility, deployment, and maintenance, traveling around Mars to locate suitable areas for settlement, measuring the water content in the soil, transporting the landers, and general construction tasks.

Those lucky – or unlucky, depending on your perspective – enough to be chosen for this unique interplanetary reality trip will travel to Mars on board the Mars Transit Vehicle, a compact space station comprising a transit living module and a lander. The transit living module will be home to the astronauts during their seven-month trip. Once in orbit around Mars, they will enter the lander and disconnect from the module, which will remain in space, orbiting the Sun.

Given this will be a reality TV show, the communications system will be particularly important. The Mars One communications mission architecture features an aerostationary satellite (the equivalent of a geostationary satellite, meaning it is always in the same place in the sky over Mars, receiving data from the settlement and transmitting them to Earth) over the Mars settlement, one in orbit around the Sun, and ground stations on Earth.

Given how long a trip to and from Mars will likely take, the Mars One spin on manned missions to Mars has the potential to serve up plenty of personal drama for those watching back on Earth. Let's just hope the ratings stay high to avoid the show being canceled, in which case, those first Martian visitors will definitely be stranded forever.

"I think we'll be able to send, probably, the first people to Mars in roughly 12 to 15 years. That's my estimate."

Elon Musk

Whether SpaceX achieves the first Mars landing alone or whether Mars One gets there with SpaceX hardware, perhaps the only surefire way to realize such a trip is a private–public partnership. On its own, government-funded science tends not to have a goal beyond pure exploration, but the entrepreneurial spirit of big companies such as SpaceX may be crucial to realizing such big ventures. A private company that's just out to make a quick profit probably won't succeed in getting to Mars, but combining the drive for exploration with an entrepreneurial spirit is what drives the world forward.

Could a Red Dragon Mars mission happen, whether it be a Mars One reality television show or a standalone SpaceX mission? Well, first of all, there are two big reasons a trip to Mars is so expensive: escaping Earth's gravity and then keeping the astronauts alive during the long trip to Mars. So, the question is how does commercial enterprise go about reducing the cost of leaving LEO and sustaining life support for such a long journey? The good news is that SpaceX has dramatically reduced the cost of getting off our planet. So, if SpaceX can make getting into LEO both cheap and reliable, and increase the frequency of flights into LEO, then they will have helped overcome one of the big hurdles of the "getting to Mars" problem. But evolving its launchers into a super-heavy-lift family that can serve as a basis for a Mars-capable architecture to LEO is only part of the problem. What about getting to the Red Planet? Well, for the transition from Earth to Mars, SpaceX believes nuclear thermal is the preferred propulsion for the piloted phase of the mission, while solar-electric power could be used to transport supplies. The nuclear option, discussed briefly in Chapter 3, is an integral part of SpaceX's exploration vision and it's a part that makes perfect sense. First, using chemical rockets makes no sense at all because a Mars mission would need as many as 15 chemical-powered vehicles to perform the same mission as two nuclear-powered Earth-departure vehicles. Secondly, the nuclear-powered option would be capable of faster transit times. Thirdly, the technology for the nuclear option could be derived from the Nuclear

Engine for Rocket Vehicle Application (NERVA) program, which included more than 17 hours of hot-fire tests and restarts. Fourthly, SpaceX could leverage Russian technology based on the Russian proposals made in 2009 to revitalize its earlier nuclear-propulsion work. Fifthly, it's not as if this technology is new – it's been around for 40 years and, 40 years ago, the technology was still the best option for a Mars mission and still better than today's chemical rockets. Since the nuclear option is such a game-changer, any company serious about traveling to Mars would be crazy not to give it some serious thought.

In SpaceX's Mars concept of operations, a fleet of tugs powered by clusters of solar-electric-powered thrusters would each carry about four metric tonnes of payload in a 390-day round trip. To land on Mars, SpaceX proposes a liquid-oxygen (LOX)/methane-powered propulsion system capable of landing a 35-tonne payload. Powering the LEO vehicles would be SpaceX's Merlin 2 engine, capable of a projected 1.7 million pounds of thrust at sea level and 1.92 million pounds of thrust in a vacuum. This engine would power the Falcon X and XX heavy launch vehicles – three Merlin 2's would power the first stage of Falcon X, a monster of a vehicle capable of placing 38,000 kilograms into LEO. Building on Falcon X would be Falcon X Heavy, which would use nine engines clustered in three cores which, collectively, would generate 10.8 million pounds of thrust at launch and carry 125,000 kilograms into orbit. Developing Falcon X Heavy would eventually result in the creation of Falcon XX, a behemoth capable of lifting 140,000 kilograms into LEO.

But what if SpaceX can't take the next step beyond LEO? Well, there are some who believe that bold missions such as those planned to the Red Planet should be NASA's business. These observers suggest that once commercial enterprise takes the burden of getting astronauts into LEO off of NASA, the agency will be free to spend its time and resources on bold missions such as getting people to Mars. After all, the agency has been putting people into LEO for 50 years now so, with commercial enterprise taking over the job, there's no reason for NASA to keep doing it. On one level, it's an argument that makes sense, the rationale being that government can help buy down the risk and prove the technology so that at some point in the future, commercial enterprise can step in and take over just as they're slowly doing in the LEO arena. But what if NASA can't get the funding for a Mars mission or what if commercial enterprise wants to go it alone? Could private enterprise be the one to get us there in the next decade? The signs are promising. The proven capability of SpaceX, combined with Musk's ambition, may represent the best bet for ultimately putting boot prints on Mars and set a precedent for future Mars travel and manned exploration. The future of space exploration seems to be in the hands of ambitious companies like SpaceX, whose sole focus is space development and exploration. Hopefully, SpaceX can achieve their stated goal of setting a man on Mars in the next 10–20 years. If successful, the company may just start a new era of private, competitively priced space travel and exploration.

9

The next great adventure: The route to commercializing low Earth orbit

In 2011, when the Space Shuttle *Atlantis* made its final flight, America marked the end of an era of publicly funded space exploration. But, while the iconic winged spacecraft that for 30 years symbolized US dominance in space was retired to various museums, the journey didn't stop with *Atlantis*. A year later, following the successful flight of Dragon, SpaceX picked up the baton and is now carrying NASA's legacy forward – thanks to Elon Musk. While SpaceX benefitted from sizeable chunks of NASA funding, Musk still spent a sizeable share of his private fortune developing the Falcon family of launchers and Dragon, and seeing his company through some tight spots along the way, among them the May 2012 Dragon mission. Dragon's success was undoubtedly a watermark in spaceflight history, ushering in the beginning of a new era when private companies, rather than governments, will challenge the "final frontier".

I usually don't watch unmanned launches. After all, it's not as if NASA hasn't figured out all of this before. But Dragon's launch had potential historic significance, so I decided to watch Dragon's first launch attempt on NASA-TV. Although former NASA astronaut, Dr Garrett Reisman, and now senior engineer at SpaceX, assured those watching there was always the possibility of a problem delaying the launch, the SpaceX team was upbeat, believing the Falcon 9 rocket would be ready to go on May 19th, 2012. But, as so often happens with events involving complex systems, the clock ticked to zero and the rocket didn't move. There was no lift-off because a problem in an engine pressure valve had automatically shut down the launch sequence. Although there was disappointment and frustration, Dragon's last-second launch abort reminded everyone that spaceflight is still hard. Really hard. Launch aborts occur. Rockets blow up. Flight profiles go out of whack. Sensors inexplicably fail. Trajectories are missed. Engines shut down for no reason. Let's face it, spaceflight is fraught with risk, and Musk is the first to accept that. Musk is also the first to acknowledge that you don't build a rocket without studying some history and, as anyone familiar with space history knows, missions don't always go to plan. Take the dramatic flight of Scott Carpenter, a mission that preceded Dragon's historic flight by almost exactly 50 years.

On May 24th, 1962, Carpenter was in the final hour of his five-hour mission, NASA's fourth manned flight. His mission included hours of scientific experiments, which distracted him from the re-entry checklist. Making matters worse, Carpenter's Aurora 7 capsule was critically short on fuel and, as the capsule passed over the Hawaii ground station with just three minutes of contact remaining, Mission Controllers weren't sure whether the vehicle was pointing in the right direction. If Carpenter's capsule wasn't pointing in the right direction for re-entry, it would bounce off the atmosphere or burn up. Either way, Carpenter would be dead.

As Carpenter checked his pre-retro-sequence tasks (equipment stowage, establishing retro-attitude, etc.), he noticed he had a problem reverting to autopilot, reporting "I have an ASCS problem here". The ASCS would not hold the $34°$ pitch and $0°$ yaw attitude required at retrofire. Making matters worse, the pitch horizon scanner was showing a false read of the horizon and had jerked the spacecraft off proper retro-attitude in pitch and yaw. Fortunately, Carpenter manually established proper pitch attitude and resumed control of the spacecraft, but he didn't have enough time to bring the spacecraft to the required $0°$ in yaw (post-flight analysis showed that, at retrofire, the capsule was yawed to the right about $25°$). The yaw issue was compounded when the retrorockets failed to fire automatically, so Carpenter had to fire them manually, but the three seconds it took to fire them, combined with under-thrusting retrorockets (and the yaw error), resulted in a 400-kilometer overshoot of the planned landing zone, with no recovery forces in sight. Because Carpenter landed outside NASA's line-of-sight radio range, he was unable to communicate with the Cape. In the 55 minutes it took to locate the Mercury astronaut, news networks speculated Carpenter had been lost, fueling the public's mounting concern for the astronaut's safety. After egressing his spacecraft, Carpenter, resting in a life raft, was joined about an hour after splashdown by Air Force para-rescue divers before being flown to Grand Turk Island for two days of debriefings, additional medical tests, and an afternoon of scuba-diving. The overshoot, the hour-long silence, and public concern for Carpenter's well-being combined to make the flight of Aurora 7 one of the more dramatic US spaceflights.

Even 50 years after Carpenter's flight, getting to space and back is still fraught with challenges. Soon, the Dragon capsule will be approved for human spaceflight and the stakes will be even higher. SpaceX has only succeeded because it learned the lessons of the past – from the failures and near disasters like Carpenter's mission.

When the Falcon 9/Dragon combination finally did launch, there was the predictable frenzy of excitement; virtual crowds tweeted about the new era in manned spaceflight, which was a little strange because SpaceX hadn't done anything particularly new – it had just done it differently. Still, the Dragon flight was a great save for the company who, at the beginning of 2012, might have begun to regret their slogan that 2012 would be the "Year of the Dragon" – a reference to the current cycle of the Chinese calendar and the upcoming tests of the company's Dragon vehicle. As the testing of the Dragon capsule delayed the launch from its original date of February 7th, 2012, it looked as though the Year of the Dragon was starting a bit late. But Dragon's May and October visits to the International Space Station (ISS) more than made good on the company's "Year of the Dragon" prediction.

Added to these ground-breaking flights were the successful test of SpaceX's Super-Draco engines for the Dragon spacecraft, the continued development of Falcon Heavy, and of course that US$440 million CCiCap funding.

While Dragon's docking and rendezvousing with the ISS represented nothing short of a historic event in the annals of spaceflight, the flight was just another in a series of SpaceX steps towards realizing a much grander goal, namely the colonization of the planet Mars. Musk has said he wants to see 10,000 people living on Mars in the near future – preferably more. At the centerpiece of this extravagant ambition is Dragon. At first glance, Dragon looks tailor-made for routine flights to and from low Earth orbit (LEO), but it doesn't exactly scream "Mars". At just 2.9 meters high, with a diameter of 3.6 meters and weighing 4,200 kilograms, it bears more than a passing resemblance to the Apollo Command Service modules that reached the Moon. It even shares a similar configuration, with a return capsule forward and a support module containing engines and support systems aft. Diminutive it may look, but Dragon and its Red Dragon Mars-bound variant (Figure 9.1) represent a part of potentially the greatest revolution in spaceflight since Sputnik was sent into orbit in 1957. For decades, spaceflight has been a government monopoly. Even when private companies finally started going into space in the 1990s, it was only as providers of launch services to send commercial and government satellites into orbit. Now, we have more than half a dozen companies working on sending tourists and scientists on suborbital flights while others are testing orbital habitats and developing plans for lunar bases. Meanwhile, the US government has ceded manned flights to LEO to private companies, thereby freeing up NASA to focus on asteroid missions and preparing to send crews on deep-space missions.

Leading the charge is SpaceX, whose Dragon has already proved its ISS docking and rendezvous capabilities. But, for Elon Musk, the ISS was never more than a stopping-off point en route to reaching Mars. In fact, just about every phase in the development of the Dragon and Falcon rockets has not been about fulfilling the next mission, but has been targeted on the steps ahead. It's a very different approach from the one NASA adopted during the early days of the manned space program. Back in

9.1 If SpaceX can continue their success, a Red Dragon might land on Mars by the mid-2020s. Courtesy: SpaceX

the 1960s, the agency was launching new rocket models like Detroit did cars, and almost as often. And, just like no one would ever confuse a Ford Falcon with a Chevrolet Corsair, no one would ever mistake a Gemini capsule for a Mercury or for an Apollo. Each vehicle was distinct, built for specific missions.

Fast forward five decades to the era of commercial spaceflight and the approach is very different. Musk's tactic is one where SpaceX uses parallel development with the goal of building launchers and capsules that fulfill client missions but are also incremental steps in a longer development program. A similar strategy was employed by the US Navy in its Polaris (Figure 9.2) submarine program of the 1960s, when submarines were designed to carry the missiles available 10 years into the future as well as those ready to be operational at the time the submarine was built.

So it is with SpaceX's launchers and capsules; Falcon 1 was built as a stepping stone to the much larger Falcon 9, which in turn was built looking forward to the Falcon Heavy and Super Heavy variants. The same strategy has been applied to the Dragon capsule with its cargo, passenger, and long-duration options. Another part of the SpaceX approach is to benefit as much as possible from the huge stores of NASA's spaceflight data, which have provided the company with not only a tremendous technological advantage, but also valuable lessons of examples of how and how not to build spacecraft. Another key factor in their success is the motivation to deliver payloads to orbit as fast as possible, as often as possible, charging as little

9.2 US Navy Polaris submarine. The development of SpaceX's launchers uses a similar path. Courtesy: US Navy

as possible. This means paying close attention to the bottom line and striving for maximum efficiency; it's the only way to make money.

Musk's plans to achieve his goal are already evident: take proven rocket designs, simplify them, and then streamline them as much as possible to build them quickly and cheaply. Since SpaceX's chief rocket designer doesn't have to worry about distributing jobs geographically, it's a strategy that will likely pay off; one of the problems NASA had – and still has – when it comes to keeping costs down is that the space program has always been seen by politicians as a way of creating jobs, especially in poorer regions of the US. It's one of the reasons why NASA's launch facilities are in Florida but Mission Control is in Texas. It also explains why there are space centers in places like Alabama. With such a distribution of people and facilities, it isn't surprising that it cost north of half a billion dollars to launch a Space Shuttle! SpaceX, on the other hand, uses as few people as possible in as central a location as possible building rockets in as few steps as possible.

Being able to draw on NASA's expertise has paid off in spades for SpaceX, as is evident in Dragon's design. The capsule's shape is derived from the Apollo capsule and utilizes similar slightly steerable aerodynamic characteristics during re-entry. Dragon's heat shield also shares Apollo heritage, except the SpaceX version is reusable whereas the ablative Apollo version wasn't. The capsule's escape system is also an Apollo legacy item, although this, in common with the heat shield, is being improved upon.

Accessing NASA's vast experience has definitely allowed SpaceX to not only save time, but also to develop components cheaply, although saving money is only part of Musk's philosophy. Musk believes that to truly make spaceflight cheap, you need to make the flight components reliable, and that means *reusable*. The logic is simple: simple systems become more reliable and reusable, reliable systems become more simple and reusable, and reusable systems become more simple and reliable. Let's face it, one of the most frustrating aspects of spaceflight over the last few decades is that it is such a wasteful enterprise: thousands of man-hours are spent building these wonderful machines that get used once and then end up either burning up in the atmosphere, sinking to the bottom of the ocean, or gathering dust in a museum. The only partially reusable system that had any success was the Space Shuttle, but this spacecraft was a rat's nest of labyrinthine complexity. Musk wants to avoid the pitfalls of the Shuttle and envisions the Dragon system, capsule, and booster stages as having the ability to fly back to base on their own. This means that SpaceX will get its gear back and it will have a space system that is truly flight-tested.

Of course, one of the problems with spacecraft that have landing systems is that these systems add weight, which means less payload can be carried. To compensate for the added weight of the landing system, you need more powerful engines so, just as the Dragon capsule needs the SuperDraco rockets if it's going to land under its own power, the Falcon boosters need larger engines. It just so happens that SpaceX is already working on the problem.

GRASSHOPPER

Even as SpaceX became the first private company to a provide cargo delivery service to the ISS – a mission that was furthered in October 2012 as the company launched its second trip to the station – it was already at work on its next giant leap: a reusable orbital spaceship. An important element in SpaceX's reusability goal is the Grasshopper program, a part of which is the "hopping" Falcon 9 rocket (Figure 9.3). The system was tested in October 2012 and consists of a standard first stage with a single Merlin 1D engine in its tail and landing legs added on. The initial test hop was all of two meters, but it was the first step in the company's reusability plan that will give its boosters the ability to launch payloads into orbit and then land robotically so they can be refueled and launched again. Instead of plunging back to Earth and being destroyed, each of Falcon 9's two stages would fly back to Earth in a controlled maneuver, using reserve propellant to make a gentle touchdown on retractable legs.

If successful, SpaceX could bring airline-style operations into LEO and

9.3 SpaceX's "hopping" rocket. Courtesy: SpaceX

significantly reduce the exorbitant cost of leaving the planet. A LEO reusable system would be nothing short of a game-changer in an industry whose costs are difficult to fathom by the average person in the street, so let's put it into some perspective: imagine if the airline industry threw away an airliner every time it flew across the Pacific and you get some idea of the state of transportation to LEO – small wonder that space tourists have to pay upwards of US$30 million a flight! But, if SpaceX can spread the cost of a spacecraft over several flights, access to LEO could become much, much cheaper, perhaps even opening the door to routine and reliable access to space. If SpaceX were able to achieve that, then manned spaceflight takes on a whole new look and it opens the door to a whole lot of new users whose business plans suddenly become practical. An indication of what might be possible can already be seen in the suborbital spaceflight arena, where reusable vehicles in development are attracting scientists and space tourists paying anywhere from US$95,000 for a ride on XCOR's Lynx (Figure 9.4) to US$200,000 on Virgin Galactic and Scaled Composites' SpaceShipTwo. Remember, until Scaled Composites proved the feasibility of reusable vehicles in 2004, the market simply didn't exist.

Grasshopper may be on the right track to realizing reusable flight but, for the system to revolutionize spaceflight, it will have to be cost-effective and achieve shorter turn-around times – in the order of a few days rather than weeks. Whatever happens in the development of Grasshopper, the system is in stark contrast to the

9.4 XCOR's suborbital Lynx spaceplane. Courtesy: XCOR

LEO vision put forward by Congress. Managed by NASA, but designed by Congress, the heavy-lift Space Launch System (SLS) – aka the Senate Launch System – is a monster rocket touted as the government vehicle that will launch astronauts to LEO some time in the 2020s. Given the history of government-designed spacecraft, it is likely the first manned launch of the SLS will be later rather than sooner, but will Grasshopper be operational any earlier?

As described in SpaceX's September 2012 FAA launch application, Grasshopper (see Appendix V for latest developments) will be used to test reusable vertical take-off and landing flight regimes, will consist of a Falcon 9 first stage powered by a single Merlin 1D engine, and be equipped with a landing structure comprising four steel legs. The proposed range of testing will last three years, beginning with flights to 75 meters and culminating in flights to a maximum altitude of 3,500 meters. By way of comparison, one of Blue Origin's vertical take-off and landing test vehicles was lost at 13,700 meters (45,000 feet) while traveling at Mach 1.2. Blue Origin's attempts (and Armadillo Aerospace's among others) at realizing a similar vehicle suggest that what SpaceX is trying to achieve will be anything but easy. One of the first hurdles will be to recover the Falcon 9 first stage, which may prove more difficult than it sounds since the vehicle's flight profile tends to hit the atmosphere in a "belly flop" position, severely damaging the first stage. One of the solutions to this problem is to restart three of its engines after stage separation to slow the first stage, thereby easing the transition into the atmosphere. If this procedure were successful, the booster would descend to a tail-first powered landing similar to that demonstrated by the elegant DC-X (Figure 9.5).

Assuming SpaceX can recover the first stage successfully, they then have to return the second stage *safely* – a more demanding task due to the much greater re-entry speeds. SpaceX plans to deal with these challenges by firing an extendable engine and adding a heat shield at the top of the second stage, followed by a powered landing using small thrusters on the perimeter of the base. And Dragon? Well, if all goes to plan, after separating from the ISS, it will return to Earth via a powered touchdown courtesy of its side-mounted pusher launch-escape system, without the aid of parachutes at all. It's an impressive proposition, but is it achievable? Regardless of whether they can prove the system, Dragon will retain its parachutes as a safety backup to the landing thrusters. And, if SpaceX does succeed, the key to that success will likely reflect the company's unique corporate culture and deliberate manner that enable it to follow a plan allowing it to pursue reusability through the normal course of its operations by means of a parallel development effort. Another benefit of SpaceX's business plan is that as long as the company can launch its vehicles, it will be able to sequentially test the components required to reach its goal without interfering with market activity and without its customers paying for much of the effort. It's a win–win situation that is the direct result of SpaceX using a simple vehicle powered by a reusable engine of its own design that was robust enough to at least allow the possibility of an upgrade to reusability. It may take SpaceX a few years to get an expended first stage to fly back to the launch site, but those years are likely to be granted thanks to the company's unique business plan and the lack of domestic competition.

9.5 The Delta Clipper Experimental (DC-X) was an unmanned prototype of a reusable single stage to orbit launch vehicle built by McDonnell Douglas in conjunction with the US Department of Defense's Strategic Defense Initiative Organization (SDIO). Courtesy: McDonnell Douglas

Since SpaceX started in 2002, Musk has been quoted as saying that the purpose of his company was to achieve a manned mission to Mars and eventually establish a settlement on the Red Planet. As the company suffered through growing pains to launch its Falcon 1 rocket, such statements were easily dismissed. Even when the company finally achieved success on September 28th, 2008, after enduring three launch failures, the difficulties encountered in launching the rather modest Falcon 1 seemed incongruous with the grand goal of a manned Mars mission. The critics weren't silenced even as SpaceX secured contracts and produced viewgraphs showing larger and bolder rockets, including a Saturn V-sized "BFR". (The "B" stands for "big", the "R" stands for "rocket", and the "F" rhymes with "trucking"!) But, with two ISS flights under its belt, the future may be a lot closer than it used to be. The official announcement of the Falcon Heavy (Figure 9.6) marked the introduction of a comparatively affordable launch vehicle with the capacity of lifting 53 tonnes to LEO, sufficient to support a return to the Moon in two flights, and perhaps launching meaningful payloads to Mars; don't forget that the Falcon Heavy is based on the flight-proven Falcon 9 vehicle, which strongly suggests it will succeed on a technical basis.

A single successful launch of the Falcon Heavy might finally silence those critics and place Mars in the crosshairs for a company whose overriding goal is to reach the Red Planet. That event may happen sooner than many thought with the revelation that SpaceX is in discussion with NASA Ames Research Center (ARC) for a 2018 "Red Dragon" Discovery-class mission in which the company would utilize a Falcon Heavy to launch a Dragon capsule to Mars. The capsule would be equipped with a drill to dig below the Martian surface to search for signs of life.

After completing an unmanned trip to Mars, the next logical step would be to send astronauts there, but how realistic is that? The two cargo flights to the ISS and the introduction of the Falcon Heavy are nothing short of revolutionary. But, if, over the next few years, SpaceX can transition the Falcon to a fully reusable launch vehicle, then the entire arena of space exploration would have to be recalibrated. Think about it: with the advent of a fully reusable Falcon family of rockets, an unforeseen scale of space exploration becomes not just more affordable, but inevitable, and a permanent human presence on Mars finally becomes within practical reach. While Musk's vision of thousands of Red Planet settlers will have to wait for new, larger rockets, SpaceX has a plan for that as well, beginning with a staged combustion engine it wants to begin building in 2013. Let's put this possible future into perspective: the US government is about to embark on a decade of development of the SLS (Figure 9.7) – a program that will consume the equivalent cost of 144 Falcon Heavy flights at 53 tonnes each into a single 70-tonne launch. With a planned launch rate of one per year, and the 130-tonne super-heavy SLS version expected no earlier than 2032 and sporting a price tag north of US$40 billion, I would put my money on SpaceX achieving reusability with the Falcon sooner than the Senate can achieve LEO with its monster rocket.

For those who, like Musk and his SpaceX employees, believe in the value of a permanently expanding human future in space, the realization and implementation of a fully reusable launch system are, and always have been, key to achieving that

9.6 Falcon Heavy. Courtesy: SpaceX

9.7 NASA's Space Launch System. The first SLS mission – Exploration Mission 1 – in 2017 will launch an uncrewed Orion spacecraft to demonstrate the integrated system performance of the SLS rocket and spacecraft prior to a crewed flight. Courtesy: NASA

dream. It's a dream that has been a long time coming and SpaceX is still a few years – perhaps longer – from achieving it. But, with no companies in the game of pursuing reusability, you have to give credit to Musk for providing a credible path to achieving it.

Innovation doesn't stop with reusability, however. Among many of its projects, SpaceX has decided to fit Dragon with a solar array – something that has never been done with an American manned spacecraft. Until Dragon came along, spacecraft relied on fuel cells, which have the drawback of only being able to operate as long as they have fuel! Thanks to Dragon's solar panels, the vehicle will be able to remain on orbit much longer – up to a year for the cargo version. Dragon's solar array is just one of many examples of how SpaceX is revamping the space industry, but the company still has its critics. Some speculate the company will find it difficult to match the quality control that is second nature to an organization like NASA. Others insist that a manned mission to Mars is a non-starter because there is simply no reason to go. These and other critiques, however, are missing the point, and the point is this: with all the problems plaguing the Russian space agency and the need to keep the ISS stocked and its crews replaced, the short-term success of Dragon will be a high priority in the months and years to come. And, if all goes well, it may open the door for more commercial US spacecraft to enter the commercial spaceflight market. It may also free up resources for grander missions such as trips to Mars. It may open up space to tourism. It may do all or none of these – we just don't know, but at least there is one company giving commercial spaceflight a fighting chance.

What we do know is that, thanks to SpaceX, NASA will shortly have a vehicle capable of serving as a lifeboat at the ISS in the event of an emergency necessitating evacuation. More importantly, the agency will no longer be held hostage to the financial whims of the Russians and an aging spacecraft that has erratic and unpredictable re-entry reliability. Furthermore, Dragon can support the return of seven crewmembers, and not just three, as Soyuz does, which means the crew can be increased without safety concerns. Having Dragon docked at the ISS also means the station has a functional ambulance capability which could return a sick or injured astronaut to Earth without all the crew having to abandon the station. Lastly, the agency may soon have a way of getting astronauts to the ISS – a capability it didn't have in 2011 when the Russians experienced one launch failure after another, prompting the agency to consider abandoning the station. The Russian launch failure was a pivotal event in the history of the ISS because it underscored the reliance the US had upon Soyuz, so it's worth revisiting.

On August 24th, 2011, an unmanned Russian cargo ship carrying supplies for astronauts on the ISS suffered a major malfunction after launch and crashed. The Progress 44 cargo ship had blasted off atop a Soyuz U rocket from Baikonur Cosmodrome in Kazakhstan but, less than six minutes into flight, shortly after the third stage was ignited, the vehicle commanded an engine shutdown. The capsule had been packed with almost three tonnes of food, fuel, and supplies for the six astronauts on board the ISS. The crash was unusual because the Progress vehicle had a long track record of reliability, having serviced the ISS since the first crew occupied the station in 2001. No one was more shocked than the Russians, whose Soyuz workhorse had logged 745 successful launches against just 21 failures since the vehicle came into service in 1966. Had a crew been on board, they probably would have been able to execute a successful abort but, given the small margin of error for future flights, the Soyuz vehicle was grounded until further notice. While there was no risk to the crew, who had plenty of supplies, the crash threw a spanner in the works of the planned crew rotation because a fresh crew of three was supposed to have been launched on September 22nd to relieve half of the station occupants. If that flight didn't happen, the crew could only stay on board for so long. While two Soyuz spacecraft were docked at the station, serving as lifeboats *in extremis*, safety rules prohibited the vehicles from remaining in space for more than 200 days, because their batteries could lose power (this won't be a problem when Dragon is fitted with its solar array) and corrosive thruster fuel could degrade rubberized seals. And, with the clock expiring in September for one of the capsules and early December for the other, without a new crew, the astronauts were faced with the bleak prospect of switching off the lights, shutting the hatch, and vacating the station.

The crash said a lot about NASA's (and the US government's) long-term planning skills because the agency had known for years that the Shuttles were living on borrowed time and yet here they were with only one transport provider and no redundancy. In 2004, the Bush Administration had launched a bold return-to-the-Moon program, instructing NASA to build two new launch vehicles, a new crew capsule, *and* a lunar lander. It was the Constellation program and it didn't last long because, in 2010, NASA and the Obama administration scrapped it, deciding to outsource LEO operations to the private sector – a decision that left a space-access gap that wasn't guaranteed to be filled by a new NASA vehicle anytime soon.

While SpaceX still has to adapt the cargo vessel so it can carry crew, at least there is one homegrown vehicle close to being man-rated and, *in extremis*, NASA has a way to get to the orbiting outpost in an emergency – a capability it didn't have in 2011. While (at the time of writing) Dragon doesn't yet have a launch-abort system (LAS) (this is being developed as part of the commercial crew activities), NASA might consider it preferable to endure the higher risk of a loss of crew without a LAS than to abandon the hundred-billion-dollar station, as the agency briefly contemplated following the Progress incident. Nobody has polled the astronauts but, following the two missions to the ISS, I bet many astronauts would probably be willing to take a ride in it, LAS or no LAS, if it was a critical mission; whether NASA would allow it is another matter.

Fifty years ago, Wernher von Braun, of Saturn V fame, was the voice of the US space program. A brilliant man, it was von Braun's vision and engineering expertise that propelled the US to the Moon. Without von Braun, many doubt Kennedy's dream of reaching the Moon before the end of 1969 would have been achieved because, without visionaries like von Braun, projects die on the vine. Fortunately, in Elon Musk, the commercial spaceflight industry has its von Braun, capable of creating rockets and daring to dream big, so it was only fitting that Musk was awarded the Wernher Von Braun Memorial Award[1] in 2009. Since the Space Shuttle *Atlantis* made its final landing, there have been innumerable op-ed pieces griping and grousing that NASA's temporary pause in the US government's manned spaceflight capability program means that America has somehow abandoned space and that the US has taken a step back from national greatness that embodied the Apollo era. But, with the successes of SpaceX, it's hard to argue that the country has abandoned space and, with the company's launch-cost-reduction program, the cost of space access is going down. The first flight of the expendable Falcon Heavy will demonstrate costs of about US$1000 a pound in 2013 and, if bulk launches of a reusable Falcon Heavy can be realized, launch costs may fall to about US$300 per pound by 2020. With such low orbital cost, the US can once again start thinking about venturing beyond LEO and begin realizing the visions von Braun dreamt of half a century ago.

COMMERCIALIZING LEO

Of course, SpaceX isn't the only company trying to get into the space business. Sir Richard Branson and his Virgin Galactic project aim to establish a suborbital passenger service that will give people the opportunity to become civilian astronauts. Robert Bigelow, founder of Bigelow Aerospace, hopes to rent out his inflatable habitats to governments and industry. These and SpaceX's competitors now represent a worldwide burgeoning industry of companies devoted to commercial spaceflight – a phenomenon that is turning the space industry on its head to the extent that, one day, NASA and the European Space Agency (ESA) may find themselves having to compete with private industry for the very best engineers.

And, with companies like SpaceX poised to take over portions of US advancement in space, some observers are beginning to ask the obvious questions: can the task of space exploration be left to private industries to develop and are private industries even capable of tackling space development safely and efficiently without government oversight and funding? Having witnessed the recent successes of SpaceX, many observers would answer in the affirmative but, before commercial space companies start planning for missions to Mars, there needs to be some evidence that SpaceX and its competitors are capable. Now, obviously, we can't point to past examples of private industries developing space, but there are

---

[1]   The award is presented in odd-numbered years to recognize excellence in the management of and leadership in a significant space-related project. The award was originally proposed in 1992 by National Space Society Awards Committee member Frederick I. Ordway III, a close associate of von Braun. The award consists of two rockets standing on an inlay of black Italian granite. The rockets are of the Von Braun "Ferry Rocket" (seen on Disney's TV episode "Man into Space" in the mid-1950s) and the Saturn V designed and built by von Braun and his Apollo team.

precedents in the field of exploration. For example, the University of Washington and Emory University completed a study that compared public and private expeditions to the North Pole between 1818 and 1909. Back in those days, Arctic and Antarctic exploration was funded either by government largesse or from the pockets of wealthy financiers. The study is an interesting one because Arctic exploration and space exploration share many common features. First there are the inhospitable conditions that test even the most advanced technologies. Then there are the myriad unknown and unforeseeable dangers. Thanks to the meticulous diaries kept by polar explorers and detailed ship logs, researchers were able to assess variables such as crew size, vessel tonnage, past experience of captains, and the number of deaths on the expedition. After a rigorous assessment of all those factors, they came up with the following conclusion:

> "Most major Arctic discoveries were made by private expeditions. Most tragedies were publicly funded. Public expeditions were better funded than their private counterparts yet lost more ships, experienced poorer crew health, and had more men die. Public expeditions' poor performance is not attributable to differences in objectives, available technologies, or country of origin. Rather, it reflects a tendency toward poor leadership structures, slow adaptation to new information, and perverse incentives."
>
> Quote from "Public versus Private Initiative in Arctic
> Exploration: The Effects of Incentives and Organizational
> Structure", *Journal of Political Economy*, **109**(1) (2001)

So, what does this have to do with the question of whether or not commercial companies can tackle space exploration? Well, according to the study's findings, not only would the private sector be capable, but it would probably do better than publicly funded competitors. That's because the study concluded that one of the problems with publicly funded expeditions is that they are slow to adapt to new information. Such a statement is still true today, as evidenced by the fact that NASA used the Shuttles for 30 years and then, instead of improving them, put them in museums. The same argument can be applied to the use of Soyuz – 1960s' technology still being used five decades later. Another liability of a publicly funded venture, as the study points out, is the problem of incentives: publically funded expeditions have a set budget that pays people the same amount regardless of the outcome – whether you do an outstanding job or whether it's decidedly sub-par, you get paid the same. Not so with a private company, which is positioned to make profits if the expedition goes well; in this case, better performance means greater gains, which is exactly how incentives are supposed to work.

So, will it be SpaceX leading the way? Well, there is no question that, at the time of writing, the company is on a roll, but there are still questions that need to be answered. One of these is whether the company can ramp up and maintain steady, reliable operations, *and* low prices. But, given the number of Falcon 9 and Dragon vehicles being prepared to fly, combined with the aggressive testing the Merlin engines and Falcon rockets are subjected to before launch and SpaceX's control over its own component supply, it's difficult to see the vehicles having a bad day. As Bill

Gerstenmaier, NASA's human spaceflight administrator, stated in one of many press interviews, "there is none better" than the SpaceX team.

Another question is whether SpaceX's competitors will deliver. That's difficult to say. At the time of writing, United Launch Alliance is facing spiraling costs and companies such as the Orbital Sciences Corporation are several years behind. China, SpaceX's lowest-cost foreign competition, hasn't got a prayer of competing with SpaceX's pricing and, as far as foreign alternatives go, the only reliable launcher is the decidedly pricey Ariane. Even when it comes to emergent launch capabilities, SpaceX either has no competition (in the heavy-lift arena) or has the edge on their competition (reusability). So, assuming SpaceX doesn't suffer a major setback over the next few years and assuming another strong competitor doesn't emerge, most routes to orbit and beyond may go through SpaceX by some time in the 2020s.

But what happens when SpaceX *does* have a bad day? After all, this is a risky business and there is always a chance of a major malfunction just as there is a chance that SpaceX's supply chain or operators may get lax, with predictably disastrous results. Will Musk resign and take his altruistic spirit with him? Will SpaceX simply ramp up the prices? Or will fiscal mismanagement put the whole company at risk? If that happens, what will be the alternative for domestic and government payloads and crews? It's worth thinking about because the government's Multi Purpose Crew Vehicle (MPCV) and SLS aren't serious competition, being further behind than SpaceX's commercial competitors and not scheduled to become operational until the early 2020s (at best). It's why the CCiCap decision not to down-select to one provider was the right move because the commercial crew program needs to have multiple, well-funded competitors so that real alternatives to Dragon are developed sooner rather than later. Just in case.

On the subject of government alternatives, SLS is unlikely to survive a decade of development in an environment of flat to slightly declining funding, looming big cuts to the federal discretionary budget, and multiple changes in the White House and Congress. If the government was serious about heavy lift being vital to deep-space activities, SLS funding should be reduced and redirected to a common off-the-shelf (COTS) program such as the Falcon Super Heavy perhaps. Another point to consider is if SpaceX becomes too successful. For example, if the Grasshopper full-scale test program produces a working reusable stage in the next few years and Blue Origin and the Air Force's Reusable Booster System (RBS) program loses ground in fielding operational reusable stages, the launch market could implode around SpaceX's order-of-magnitude cost advantage because SpaceX will have unprecedented pricing power and the ability to run competitors out of business. This would be great for SpaceX, but not so good for payload developers or the government. If we don't want this to happen, there needs to be a COTS program for reusability.

Appendix I

Elon Musk: Presidential commission speech

As members of the Commission are aware, the cost and reliability of access to space have barely changed since the Apollo era over three decades ago. Yet in virtually every other field of technology, we have made great strides in reducing cost and increasing capability, often in ways we did not dream existed. We have improved computing costs by a factor 10,000 or more, decoded the human genome and built the Internet. The exception to this wave of development has been space launch, but why?

My best guess at the origin of the problem relates to a breakdown of what the economist Schumpeter called "creative destruction". He postulated that the way an industry improves is that new companies enter a market with a lower price or superior product. This forces the whole market to improve. Looking at orbital launch vehicles, we see a situation where there has been no successful new entrant in four decades, apart from one firm established in the late 1980s. Moreover, there has actually never been a truly commercial development anywhere in the world that reached orbit.

To address this problem, we must create a fertile environment for new space access companies that brings to bear the same free market forces that have made our country the greatest economic power in the world. If we can create such an environment, I expect that progress in space launch costs and capability will be no less dramatic than in other technology sectors.

If you doubt that we can possibly see such progress in space access, please reflect for a moment that the Internet, originally a DARPA funded project, showed negligible growth for over two decades until private enterprise entered the picture and made it accessible to the general public. At that point, growth accelerated by more than a factor of ten. We saw Internet traffic grow by more in a few years than the sum of all growth in the prior two decades.

We are at a crucial turning point today. The vision outlined by the President is absolutely achievable within the current NASA budget and schedule, *but only by making use of new entrepreneurial companies along with the incumbents*. It cannot be achieved at all if we simply follow the old paths, which have led us to one cancelled program after another since the Space Shuttle.

What strategies are key to achieving the President's vision?

1. Increase and Extend the Use of Prizes

Offering substantial prizes for achievement in space could pay enormous dividends. We are beginning to see how powerful this can be by observing the recent DARPA Grand Challenge and the X Prize. History is replete with examples of prizes spurring great achievements, such as the Orteig Prize, famously won by Charles Lindbergh, and the Longitude prize for ocean navigation.

Few things stoke the fires of creativity and ingenuity more than competing for a prize in fair and open competition. The result is an efficient Darwinian exercise with the subjectivity and error of proposal evaluation removed. The best means of solving the problem will be found and that solution may be in a way and from a company that no-one ever expected.

One interesting option might be to parallel every NASA contract award with a prize valued at one tenth of the contract amount. If another company achieves all of the contract goals first, they receive the prize and the main contract is cancelled. At minimum, it will serve as competitive spur even after contract award.

We should strongly support and extend the proposed Centennial Prizes put forward in the recent NASA budget. No dollar spent on space research will yield greater value for the American people than those prizes.

2. Support new entrants in space launch

The most fundamental barrier to human exploration beyond low Earth orbit and hence meeting the President's vision is the cost of access to space. Here it should be noted that the cost of launch also drives the cost of spacecraft. If you are paying $5000/lb to put something in space, you will naturally pay up to $5000/lb to save weight on your spacecraft, creating a vicious cycle of cost inflation.

This problem of affordability dwarf's all others. If we do not set ourselves on the track of solving it with a constantly improving price per pound to orbit, in effect a Moore's law of space, neither the average American nor their great-grandchildren will ever see another planet. We will be forever confined to Earth and may never come to understand the true nature and wonder of the Universe.

It was precisely for this reason that I established SpaceX and set as our goal improving the cost as well as the reliability of access to space. Our first offering, called Falcon, will be the only semi-reusable orbital rocket apart from the Space Shuttle. Initially, we will deliver cargo to orbit in the form of satellites, however we believe strongly in the long term market for commercial human transportation.

As a starting point for improving the affordability problem, the Falcon is only one fifth the NASA list price of our US competitors. Moreover, we expect to decrease our prices in real, if not absolute terms every year, and will be announcing a price decrease in our Falcon I vehicle shortly.

New companies might also provide reliability levels more comparable with airline transportation. In the case of SpaceX, we believe that our second generation vehicle in particular, the Falcon V, will provide a factor of ten improvement in propulsion reliability. Falcon V will be the first US launch vehicle since the Saturn V Moon rocket that can complete its mission even if an engine fails in flight – like almost all commercial aircraft. In fact, Saturn V, which had a flawless flight record, was able to complete its mission on two occasions only because it had engine out redundancy.

My thanks for the opportunity to come before you today, and I look forward to answering any questions that you may have

ELON MUSK: SENATE TESTIMONY MAY 5, 2004

Mr. Chairman and Members of the Committee, thank you for inviting me to testify today on the future of Space Launch Vehicles and what role the private sector might play.

The past few decades have been a dark age for development of a new human space transportation system. One multi-billion dollar Government program after another has failed. In fact, they have failed even to reach the launch pad, let alone get to space. Those in the space industry, including some of my panel members, have felt the pain first hand. The public, whose hard earned money has gone to fund these developments, has felt it indirectly.

The reaction of the public has been to care less and less about space, an apathy not intrinsic to a nation of explorers, but born of poor progress, of being disappointed time and again. When America landed on the Moon, I believe we made a promise and gave people a dream. It seemed then that, given the normal course of technological evolution, someone who was not a billionaire, not an astronaut made of "The Right Stuff", but just a normal person, might one day see Earth from space. That dream is nothing but broken disappointment today. If we do not now take action different from the past, it will remain that way.

What strategies are critical to the future of space launch vehicles?

1. Increase and Extend the Use of Prizes

This is a point whose importance cannot be overstated. If I can emphasize, underscore and highlight one strategy for Congress, it is to offer prizes of meaningful scale and scope. This is a proposition where the American taxpayer cannot lose. Unlike standard contracting, where failure is often perversely rewarded with more money, failure to win a prize costs us nothing.

Offering substantial prizes for achievement in space could pay enormous dividends. We are beginning to see how powerful this can be by observing the X Prize, a prize for suborbital human transportation, which is on the verge of being won. It is a very effective use of money, as vastly more than the $10 million prize is being spent by the dozens of teams that hope to win. At least as important, however, is the spirit and vigor it has injected into the space industry and the public at large. It is currently the sole ember of hope that one day they too may travel to space.

Beyond space, as the Committee is no doubt aware, history is replete with examples of prizes spurring great achievements, such as the Orteig Prize for crossing the Atlantic nonstop by plane and the Longitude prize for ocean navigation.

Few things stoke the fires of creativity and ingenuity more than competing for a prize in fair and open competition. The result is an efficient Darwinian exercise with the subjectivity and error of proposal evaluation removed. The best means of solving the problem will be found and that solution may be in a way and from a company that no-one ever expected.

One interesting option might be to parallel every major NASA contract award with a prize valued at one tenth of the contract amount. If another company achieves all of the contract goals first, they receive the prize and the main contract is cancelled. At minimum, it will serve as competitive spur for cost plus contractors.

Some people believe that no serious company would pursue a prize. This is simply beside the point: if a prize is not won, it costs us nothing. Put prizes out there, make them of a meaningful size, and many companies will vie to win, particularly if there are a series of prizes of successively greater difficulty and value.

I recommend strongly supporting and actually substantially expanding upon the proposed Centennial Prizes put forward in the recent NASA budget. No dollar spent on space research will yield greater value for the American people than those prizes.

2. Rigorously Examine How Any Proposed New Vehicle Will Improve the Cost of Access to Space

The obvious barrier to human exploration beyond low Earth orbit is the cost of access to space. This problem of affordability dwarf's all others. If we do not set ourselves on the track of solving it with a constantly improving price per pound to orbit, in effect a Moore's law of space, neither the average American nor their great-great-grandchildren will ever see another planet. We will be forever confined to Earth and may never come to understand the true nature and wonder of the Universe. So it is critical that we thoroughly examine the probable cost of alternatives to replacing the Shuttle before embarking upon a new development. The Shuttle today costs about a factor of ten more per flight than originally projected and we don't want to be in a similar situation with its replacement.

In fact, it was precisely to improve the cost and reliability of access to space, initially for satellites and later for humans, that I established SpaceX (although some of my friends still think the real goal was to turn a large fortune into a small one). Our first offering, called Falcon I, will be the world's only semi-reusable orbital rocket apart from the Space Shuttle. Although Falcon I is a light class launch vehicle, we have already announced and sold the first flight of Falcon V, our medium class rocket. Long term plans call for development of a heavy lift product and even a super-heavy, if there is customer demand. We expect that each size increase would result in a meaningful decrease in cost per pound to orbit. For example, dollar cost per pound to orbit dropped from $4000 to $1300 between Falcon I and Falcon V. Ultimately, I believe $500 per pound or less is very achievable.

3. Ensure Fairness in Contracting

It is critical that the Government acts and is perceived to act fairly in its award of contracts. Failure to do so will have an extremely negative effect, not just on the particular company treated unfairly, but on all private capital considering entering the space launch business.

SpaceX has directly experienced this problem with the contract recently offered to Kistler Aerospace by NASA and it is worth drilling into this as a case example. Before going further, let me make clear that I and the rest of SpaceX have a high regard for NASA as a whole and have many friends & supporters within the organization. Although we are against this particular contract and believe it does not support a healthy future for American space exploration, this should be viewed as an isolated difference of opinion. As mentioned earlier, for example, we are very much in favor of the NASA Centennial Prize initiative.

For background, the approximately quarter billion dollars involved in the Kistler contract would be awarded primarily for flight demonstrations & technology showing the potential to resupply the Space Station and possibly for transportation of astronauts.

That all sounds well and good. The reason SpaceX is opposing the contract and asking the General Accounting Office to put this under the microscope is that it was awarded on a sole source, uncompeted basis to Kistler instead of undergoing a full, fair and open competition. SpaceX and other companies (Lockheed and Spacehab also raised objections) should have, but were denied the opportunity to compete on a level playing field to best serve the American taxpayer. Please note that this is a case where SpaceX is only asking for a fair shot to meet the objectives, not demanding to win the contract.

The sole source award to Kistler is mystifying given that the company has been bankrupt since July of last year, demonstrating less than stellar business execution (if a pun is permitted). Moreover, Kistler intends to launch from Australia using all Russian engines, raising some question as to why this warrants expenditure of American tax dollars.

Now, although we feel strongly to the contrary, it is possible that NASA has made the right decision in this case. However, does awarding a sole source contract to a bankrupt company over the objections of others sound like a fair decision? Common sense suggests the answer. Whether Kistler does or does not ultimately deserve to win this contract, it should never have been awarded without full competition.

Again, thank you for inviting me to testify before you today.

Appendix II

Applying for a launch and re-entry license

If you're planning on launching a rocket, you can't just find a suitable launch pad and start the countdown. As with most dangerous activities, if you want to launch a rocket, you need permission; in this case, permission comes in the form of a launch and/or re-entry license. Filing the request for such a license requires a lot of paperwork, which is submitted to the Federal Aviation Administration (FAA). Assuming you have all your paperwork in order, the FAA will issue you a license which will identify, by name or mission, each activity authorized under the license. Your authorization to operate terminates when you complete all launches or re-entries authorized by the license or the expiration date stated in the license, whichever occurs first.

Now, most people reading this book will never apply for a launch and re-entry license, so the following steps are purely for interest.

The first step is the pre-application consultation, which means attending various meetings to allow the prospective applicant to familiarize the FAA with its proposal and the FAA to familiarize the prospective applicant with the licensing process. It also provides a potential applicant with an opportunity to identify any unique aspects of its proposal and develop a schedule for submitting an application. Before starting this step, it's helpful to become familiar with the application procedures and the various statutes and regulations governing commercial spaceflight, such as the Commercial Space Launch Act and the *Licensing and Safety Requirements for Launch* publication. It's also worthwhile reading the various advisories that apply to the license-application procedures, such as those dealing with policy review and approval, safety review and approval, financial responsibility, and environmental reviews.

Once you've familiarized yourself with all the paperwork and launch and re-entry jargon, you have to obtain policy and safety approvals from the FAA, which means you have to apply for the approvals in advance of submitting a complete license application using the application procedures. Once you've done that, a payload re-entry determination has to be made, after which the FAA conducts a review. If your payload happens to be human, then you have to fill in a lot more forms and demonstrate compliance with a lot more requirements than if you were just planning on launching an unmanned payload.

Once all the paperwork is submitted, the FAA conducts a safety review to determine whether you – the applicant – is capable of re-entering a re-entry vehicle and payload, if any, to a designated re-entry site without jeopardizing public health and safety and the safety of property. If the FAA approves, then you are issued the license. Once you receive your launch or re-entry license, it's worth knowing what it entitles you to. In short, an operator license authorizes you to conduct launches or re-entries from one launch or re-entry site within a range of operational parameters of launch or re-entry vehicles from the same family of vehicles transporting specified classes of payloads or performing specified activities. The license remains in effect for two to five years from the date it's issued.

Appendix III

SPACEX

DragonLab™
Fast track to flight.

OVERALL DRAGON™ CAPABILITIES

Dragon is a free-flying, reusable spacecraft capable of hosting pressurized and unpressurized payloads. Subsystems include propulsion, power, thermal control, environmental control, avionics, communications, thermal protection, flight software, guidance, navigation & control, entry, descent & landing, and recovery.

USES

- Highly Responsive payload hosting
- Sensors/apertures up to 3.5m diameter
- Instruments and sensor testing
- Spacecraft deployment
- Space physics and relativity experiments
- Radiation effects research
- Microgravity research
- Life science and biotech studies
- Earth sciences and observations
- Materials and space environments research
- Rendezvous and inspection
- Robotic servicing

DRAGON SPACECRAFT SYSTEM

- Fully recoverable capsule
- Trunk jettisoned prior to reentry
- 6000 kg total combined up-mass capability
- Up to 3000 kg down mass
- Payload Volume:
  - 10 m³ pressurized
  - 14 m³ unpressurized
- Mission Duration: 1 week to 2 years
- Payload Integration timeline:
  - Nominal: L-14 days
  - Late-load: T-9 hours
- Payload Return:
  - Nominal: End-of-Mission + 14 days
  - Early Access: End-of-Mission + 6 hours

TYPICAL INTEGRATION TIMELINE

			Payload Integration	Return
AI P	ICD	Fit Check		
L - 9 Mth	~ 3 Mth	L-2 Mths ~ 2 Wks	EOM 2 wk	

Launch

v.2.1

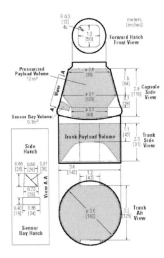

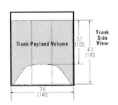

OPTIONAL TRUNK EXTENSION

SPACEX

For more information, please email us at DragonLab@spacex.com.

spacex.com

PAYLOAD SERVICES

MECHANICAL

- Specific mounting locations and environments are mission-unique
- **Pressure Vessel Interior** (pressurized, recoverable)
  - 10 m³ payload volume
  - Lab temp, pressure and RH
  - Typically Middeck Locker accommodations
  - Other mounting arrangements available
- **Sensor Bay** (unpressurized, recoverable)
  - Approx 0.1 m³ (4cu ft) volume
  - Hatch opens after orbit insertion; closes prior to reentry
  - Electrical pass-throughs into pressure vessel
- **Trunk** (unpressurized, non-recoverable)
  - 14 m³ payload volume
  - Optional trunk extension for a total of up to 4.3 m length, payload volume 34 m³

POWER

- 28 VDC & 120 VDC
- Up to 1500-2000 W average; up to 4000 W peak

THERMAL & ENVIRONMENTAL

(ref. NASA SSP 57000)

- Internal Temp: 10–46 °C
- Internal Humidity: 25–75% RH
- Internal Pressure: 13.9–14.9 psia
- Cleanliness: Visibly Clean-Sensitive (SN-C-0005)
- Pressurized: convective or cold-plate
- Unpressurized: cold-plates if required
- Payload random vibration environment:
  - Pressurized: 2.4 grms (> 100 lbm)
  - Unpressurized: 2.9 grms

TELEMETRY & COMMAND

- Payload RS-422 serial I/O, 1553, and Ethernet interfaces (all locations)
- IP addressable payload standard service
- Command uplink: 300 kbps
- Telemetry/data downlink: 300 Mbps (higher rates available)

SPACECRAFT SUBSYSTEMS

STRUCTURES AND MECHANISMS

- All Structures and Mechanisms are designed to be capable of supporting crew transportation, consistent with all relevant NASA standards and Factors of Safety
- 3 or 4 windows, 30 cm diameter
- Sensor Bay Hatch: deployable/retractable hatch mechanism which opens on orbit and closes prior to reentry
- Capsule/Trunk fluid & electrical interconnects

PROPULSION

- 12-18 Draco thrusters
- NTO/MMH hypergolic propellants
- Up to 2 fault tolerant

AVIONICS

- Dual/Quad fault tolerant Flight Computers
- Multiple generic Remote Input/Output (RIO) modules with customized complements of Personality Modules

FLIGHT SOFTWARE

- VxWorks platform
- Resides in both Flight Computers and Remote Input/Output modules
- Extensive flight heritage

COMMUNICATIONS

- Fault tolerant S-band telemetry & video transmitters
- Onboard compression & command encryption/ decryption
- Links via TDRSS and ground stations

POWER

- 2 articulated solar arrays
- Unregulated 28V main bus
- 4 redundant Lithium-Polymer batteries

GUIDANCE, NAVIGATION & CONTROL (GNC)

- Inertial Measurement Units, GPS & Star Trackers
- **Specifications:**
  - Attitude Determination: < 0.004° w.r.t. inertial frame
  - Attitude Control: < 0.012°/axis during station-keep
  - Attitude Rate: <0.02°/sec/axis during station-keep

ENVIRONMENTAL CONTROL (PRESSURE VESSEL)

- Active control of pressure & pressurization rates
- Humidity monitoring
- Air circulation and temp control

THERMAL CONTROL SYSTEM (TCS)

- Two fully redundant and independent Pumped Fluid Loops
- Radiator mounted to trunk structure

THERMAL PROTECTION (TPS)

- PICA-X primary heatshield
- Large design margins

ENTRY, DESCENT & LANDING (EDL) & RECOVERY

- Water splashdown under parachutes (off CA coast)
- Redundant Drogue and Main parachutes
- GPS/Iridium locator beacons
- Ship recovery

Appendix IV[1]

Falcon Heavy overview

Falcon Heavy, the world's most powerful rocket, represents SpaceX's entry into the heavy-lift launch vehicle category. With the ability to carry satellites or interplanetary spacecraft weighing over 53 metric tonnes (117,000 lb) to low Earth orbit (LEO), the Falcon Heavy can lift nearly twice the payload of the next closest vehicle, the US Space Shuttle, and more than twice the payload of the Delta IV Heavy.

With over 3.8 million pounds of thrust at lift-off, the Falcon Heavy will be the most capable rocket flying. By comparison, the lift-off thrust of the Falcon Heavy equals 15 Boeing 747 aircraft at full power.

THE WORLD'S HEAVY-LIFT VEHICLES

Only the Saturn V, last flown in 1973, delivered more payload to orbit than the Falcon Heavy.

Vehicle	Inclination	Orbit	Payload to LEO
Falcon Heavy	**28.5°**	**200 km**	**53,000 kg**
Space Shuttle	28.5°	200 km	24,400 kg
Delta IV Heavy	28.5°	407 km	22,980 kg
Titan IV-B	28.5°	150 km × 175 km	21,680 kg
Proton M	51.6°	200 km	21,000 kg
Ariane 5 ES	51.6°	407 km	20,000 kg
Atlas V 551	28.5°	200 km	18,810 kg
Japan H2B	30.4°	300 km	16,500 kg
China LM3B	28.5°	200 km	11,200 kg

---

[1] Appendix adapted from the Falcon Heavy Overview posted on the SpaceX website at *www.spacex.com/falcon_heavy.php.*

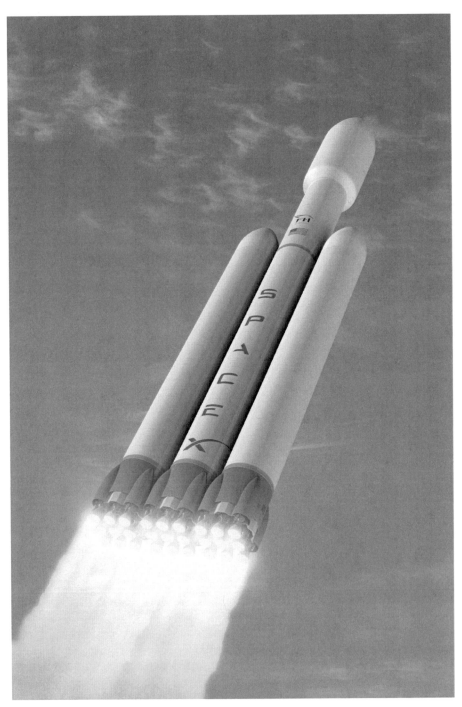

The Falcon Heavy. Courtesy: SpaceX

The Falcon Heavy's first stage will be made up of three nine-engine cores, which are used as the first stage of the SpaceX Falcon 9 launch vehicle. It will be powered by SpaceX's upgraded Merlin engines currently being tested at the SpaceX rocket development facility in McGregor, Texas. SpaceX has already designed the Falcon 9 first stage to support the additional loads of this configuration and, with common structures and engines for both Falcon 9 and the Falcon Heavy, development and operation of the Falcon Heavy will be highly cost-effective.

Falcon Heavy specifications

Mass to LEO (200 km, 28.5°)	53,000 kg
Overall length	69.2 m
Width (body)	3.6 m × 11.6 m
Width (fairing)	5.2 m
Thrust on lift-off	17 MN

HIGH RELIABILITY AND HIGH PERFORMANCE

The Falcon Heavy is designed for extreme reliability and can tolerate the failure of several engines and still complete its mission. As on commercial airliners, protective shells surround each engine to contain a worst-case situation such as fire or a chamber rupture, and prevent it from affecting the other engines and stages. A disabled engine is automatically shut down, and the remaining engines operate slightly longer to compensate for the loss without detriment to the mission.

The Falcon Heavy will be the first rocket in history to feature propellant cross-feed from the side boosters to the center core. Propellant cross-feeding leaves the center core still carrying the majority of its propellant after the side boosters separate. This makes the Falcon Heavy's performance comparable to that of a three-stage rocket, even though only the single Merlin engine on the upper stage requires ignition after lift-off, further improving both reliability and payload performance. Should cross-feed not be required for lower-mass missions, it can be easily turned off.

Anticipating potential astronaut transport needs, the Falcon Heavy is also designed to meet NASA human rating standards. The Falcon Heavy is designed to higher structural safety margins of 40% above flight loads, rather than the 25% level of other rockets, and triple redundant avionics. Despite being designed to higher structural margins than other rockets, the Falcon Heavy side booster stages have a mass ratio (full versus empty) above 30 – better than any launcher in history. By comparison, the Delta IV side boosters have a mass ratio of about 10.

FAIRING VOLUME

Below are the standard fairing dimensions for the Falcon Heavy. Dimensions are in meters and in inches inside the brackets. Custom fairings are available at incremental cost.

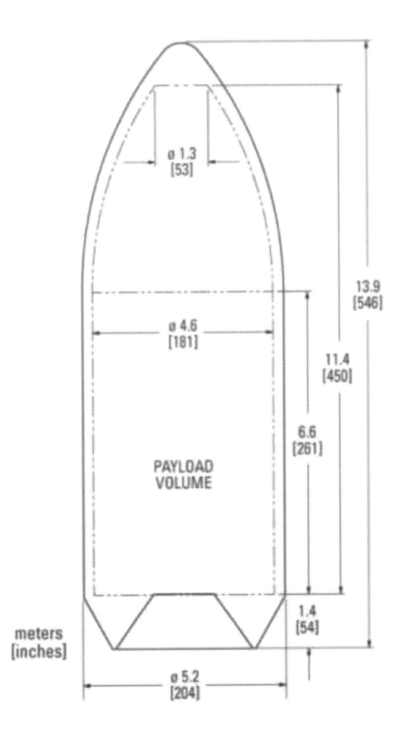

ø 1.3
[53]

13.9
[546]

ø 4.6
[181]

11.4
[450]

6.6
[261]

PAYLOAD
VOLUME

1.4
[54]

meters
[inches]

ø 5.2
[204]

LAUNCH AND PLACEMENT SERVICE PRICING

SpaceX offers open and fixed pricing for launch services based on SpaceX's standard statement of work. Additional mission assurance activities or other non-standard services are available for an additional charge. Modest discounts are available for contractually committed, multi-launch purchases.

Payload	*Price*
Up to 6.4 tonnes to geosynchronous transfer orbit (GTO)	US$83 million
Greater than 6.4 tonnes to GTO	US$128 million

Performance		
	Launch site	Cape Canaveral AFS
	Mass to LEO	53,000 kg
	Inclination	28.5°
	Mass to GTO	12,000 kg
	Inclination	27°

Appendix V

Latest developments

NOVEMBER 2012

For years, rocket engineers have talked about the realization of building a fully reusable orbital launch system. In November 2012, Space X took another step towards that goal when its Grasshopper, a 10-storey vertical take-off and landing (VTVL) vehicle (Figure A1), lifted more than five meters off the ground, hovered, and touched back down on the pad – all without a hitch.

A1. Grasshopper in vertical flight. Courtesy: SpaceX

Comprising a Falcon 9 rocket first stage, a Merlin 1D engine, four landing legs with hydraulic dampers, and a steel support structure, Grasshopper, if successful, will revolutionize delivery of cargo to orbit. Using expendable launch systems where you don't attempt any recovery, you only get maybe 2–3% of your lift-off weight to orbit. With reusability, every element of the launch system is used again, thereby driving down costs. But it's not easy because, to make a vehicle reusable, engineers need to strengthen the stages and add thermal protection, which, of course, adds weight, and added weight means less payload. As always, SpaceX posted video of the short flight on their website (*www.spacex.com/updates.php*). The short clip shows the vehicle launching towards a nominal first-stage main engine cut-off (MECO) and staging prior to the entire first stage rotating 180° using its reaction control system (RCS) thrusters, and then re-igniting three of the nine engines to "boost back" the near-empty stage back to the launch site.

DECEMBER 2012

The November test was followed by another the following month when Grasshopper rose to 40 meters, hovered, and then smoothly returned to the ground on its landing legs. It was by far the highest hop by Grasshopper. There are still plenty of flight tests, some of which will need to replicate the challenges facing the first stage, ahead of evaluating a test plan for the upper stage. In operational mode, the upper stage would complete its orbital insertion burn prior to spacecraft separation. Protected by a heat shield used by the Dragon spacecraft, the upper stage (after dropping off the satellite or spacecraft in orbit) will dive back to Earth prior to using four thrusters to decelerate and land on its landing legs. Eventually, the plan is for Dragon to propulsively land on terra firma, using its Dragon's Draco thrusters – that initially have a standby role as the Launch Abort System (LAS) motors during ascent.

Prior to Grasshopper's 40-meter altitude flight, the US Air Force (USAF) Space and Missile Systems Center awarded SpaceX two Evolved Expendable Launch Vehicle (EELV)-class missions: the Deep Space Climate Observatory (DSCOVR) and the Space Test Program 2 (STP-2). Planned to be launched on the Falcon launch vehicle in 2014 and 2015 respectively, the awards marked the first EELV-class missions awarded to SpaceX. The DSCOVR mission will be launched aboard a Falcon 9, while the STP-2 will be launched aboard the Falcon Heavy. Both missions fall under the Orbital/Suborbital Program-3 (OSP-3), an indefinite-delivery/ indefinite-quantity contract for the USAF Rocket Systems Launch Program:

> **DSCOVR** mission: this mission will place the DSCOVR satellite at the Sun–Earth Lagrange point L1, approximately 1,500,000 kilometers from Earth, where it will monitor Earth and space weather, providing warning of space weather events that may impact civilian and military activities on Earth.
>
> **STP-2** mission: this mission comprises two space vehicles – the Constellation Observing System for Meteorology, Ionosphere, and Climate-2 (COSMIC-2), designed to monitor climate behaviors, and the Demonstration and Science

Experiments (DSX), which will perform radiation research for the Department of Defense. To execute this mission, a Falcon Heavy will perform two orbital insertions, deploying COSMIC-2 into low Earth orbit (LEO) and the DSX into medium Earth orbit. Thanks to a secondary payload adapter, this mission will also deploy cubesats at each insertion point.

JANUARY 2013

At the end of January, SpaceX and Space Communication Ltd (Spacecom) announced an agreement to launch Spacecom's AMOS-6 satellite (which will replace AMOS-2, launched in 2003) on a Falcon 9. Under the agreement, an upgraded (featuring extended first-stage fuel tank, more powerful Merlin engines, and an altered first-stage engine configuration) Falcon 9 will insert the communications satellite into a geosynchronous transfer orbit (GTO) some time in 2015 from Cape Canaveral, Florida. The AMOS-6 satellite will be built by Israel Aerospace Industries (IAI) and will provide communication services including direct satellite home internet for Africa, the Middle East, and Europe. The AMOS-6 agreement was the latest in a long line of launch contracts for SpaceX, which signal 14 launch contracts in 2012 alone, maintaining the company's position as the world's fastest-growing launch services provider.

FEBRUARY 2013

February 25th, 2013, was all about preparation (Figures A2 and A3) for the Commercial Resupply Services (CRS)-2 flight to the International Space Station (ISS), slated for March 2013. At the end of the month, SpaceX conducted a successful static fire test of the Falcon 9 rocket.

The nine-engine test, which took place at the company's Space Launch Complex 40 at the Cape Canaveral Air Force Station, was a full launch dress rehearsal for the CRS-2, the second official cargo resupply mission under NASA's CRS contract. Earlier in the month, Dragon had been prepped (Figure A4) for launch prior to arriving for mating with Falcon 9 (Figure A5). During the static fire test, SpaceX conducted all the launch-day countdown events and fired all nine engines at full power for two seconds, while Falcon 9 was secured to the pad.

MARCH 2013

The CRS-2 mission (Figure A6) that launched (Figures A7–A9) from Cape Canaveral Air Force Station, at 10:10 a.m. EST, was the second of at least 12 missions to the ISS that SpaceX will fly for NASA under the CRS contract. While the launch proceeded smoothly, only one of the capsule's four thruster pods activated after it reached orbit, highlighting once again the challenges of orbital

A2. Rollout of Falcon 9 in advance of the CRS-2 mission. Courtesy: SpaceX

A3. Falcon 9 being prepared for the CRS-2 mission. Courtesy: SpaceX

A4. Dragon capsule fitted with its solar arrays in preparation for the CRS-2 mission. Courtesy: SpaceX

A5. Falcon 9 in the hangar. Courtesy: SpaceX

A6. CRS-2 mission patch. Courtesy: SpaceX

flight (during the CRS-1 mission, one of Falcon 9's nine Merlin engines shut down prematurely, causing a telecommunications satellite that was riding along as a secondary payload to be placed in the wrong orbit). Each pod contains a group of thrusters that are used to guide Dragon's course while in orbit. With only one thruster pod running, Dragon would have been unable to rendezvous with the ISS because NASA mission rules stated that at least three pods needed to be working. While engineers worked to troubleshoot the problem, SpaceX went ahead and deployed the arrays. Then, just before 3 p.m. EST, engineers reported that thruster pods one through four were operating nominally. All systems were back in the green and, an hour later, the necessary orbit-raising burn had been completed. Dragon was back on track. The problem was suspected to be a stuck valve or some blockage in the lines for pressurizing the thruster's oxidizer tank because cycling the valves and releasing a blast of pressurized helium had cleared the lines. Resolving the pod problem was a big – and expensive – deal because, if Dragon had not been able to hook up with the ISS, SpaceX would not have received full payment for the flight. (Under the terms of a 12-flight US$1.6 billion contract with NASA, each Dragon

A7. Falcon 9 vertical on the pad. Courtesy: SpaceX

A8. CRS-2 launch. Courtesy: SpaceX

A9. SpaceX employees watch the Falcon 9 launch at SpaceX's Hawthorne head-quarters. Courtesy: SpaceX

mission costs NASA about US$133 million, so SpaceX would have been out of pocket on the CRS-2 mission.)

Fortunately, after several hours of troubleshooting, SpaceX engineers managed to resolve the situation, fixing the problem and bringing all four pods online for an engine burn that set Dragon back on course for the ISS, although the problem added a day's delay in Dragon's rendezvous with the ISS. Except for the trouble with the pods, the launch added to SpaceX's perfect track record because it was Falcon 9's fifth consecutive success.

Dragon's cargo manifest (Panel A1) contained 1,050 kilograms of cargo, including experiments to study the growth of plants and mouse stem cells in zero-G. There were also spare parts for the ISS air-recycling system, and a research freezer for preserving biological samples. Also packed on board was fruit for the crew, fresh from an orchard owned by the father of one of SpaceX's employees.

On March 3rd, at 5:31 a.m. EST/1031, the ISS and Dragon sailed 400 kilometers over northern Ukraine. The orbital ballet ended when station commander Kevin Ford, working from a robotics station inside the outpost, grabbed Dragon with the station's robot arm (Figure A10). NASA flight controllers then stepped in to drive the capsule to its berthing port on the station's Harmony connecting node. Docking occurred at 8:44 a.m. EST.

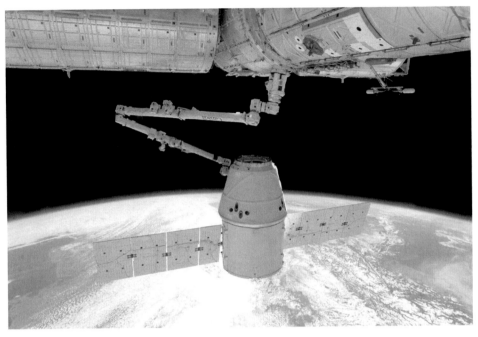

A10. Capture of Dragon by the ISS. Courtesy: SpaceX

Panel A1. Dragon CRS-2 cargo manifest: Cargo up-mass

Total cargo up-mass: 575 kg
Total mass with packaging: 677 kg

Crew supplies: 81 kg
- Personal crew care package
- Clothing and standard hygiene items
- Wet trash bags
- Crew food provisions
- Operations data files

Science hardware

Canadian Space Agency:
Microflow – 11 kg (Microflow will demonstrate a miniaturized flow cytometer that will quantify molecules and cells in blood or other body fluids in zero-gravity)

European Space Agency:
ENERGY – 11 kg (Energy = Astronauts' Energy Requirements for Long-Term Space Flight – Human Research Payload. ENERGY

measures changes in energy balance during long-term spaceflight, adaptations in the components of the Total Energy Expenditure, and it will derive an equation for the energy requirements of astronauts. For the crewmembers, ENERGY includes a special diet, urine sampling, oxygen-uptake measurements, and diet logging)

JAXA experiments: 3 kg
Bio paddles
Stem cells (the Stem Cell Study that is coordinated by JAXA examines the development of embryonic stem cells that are being flown on the Space Station. "The cells are launched frozen and after returning to Earth are microinjected into mouse-8-cell embryos in order to analyze the influence of the space environment on the development and growth of adult mice." (JAXA document))

NASA experiments: 323 kg
GLACIER – One Active (Powered) and One Passive (The Glacier (General Laboratory Active Cryogenic on ISS Experiment Refrigeration) is an experiment storage freezer which is used to keep reagents and samples in a controlled environment. It is divided into four separate compartments; these can be operated at different temperatures. Currently temperatures of $+4°$, $-26°C$, and $-80°C$ are used aboard the ISS)
BRIC – Biological Research In Canisters (BRIC examines the effects of spaceflight on small specimens that are flown inside experiment storage containers)
Cell Bio Tech Demo – Cell Biology Technology Demonstration (the demonstration will conduct on-orbit validation testing of an ISS incubator for cell culture experiments and biotechnology on the ISS. Experiments performed include studies of cell and molecular biology)
Nanoracks (NanoRacks are small experiment racks that can be placed within ISS facilities to host a number of different autonomous, self-contained experiments that can be flown quickly and inexpensively, enabling students to have payloads flown to the ISS)
CGBA – Commercial Generic Bioprocessing Apparatus Hardware
CSLM-3 – Coarsening in Solid Liquid Mixtures-3 (Materials Science Investigation that looks at the growth and solidification processes in lead–tin solid–liquid mixtures that contain traces of tin branch-like structures)
FCF supplies – FCF = Fluids & Combustion Facility
MSG supplies – MSG – Microgravity Science Glovebox
Seedling Growth ("Seedling Growth-2 is the second part of the Seedling Growth Experiment series and uses the plant *Arabidopsis thaliana* to determine the effects of gravity on cellular signaling mechanisms

of phototropism and to investigate cell growth and proliferation responses to light stimulation in microgravity conditions." (NASA ISS Program Office))

Wet Lab (The Wet Lab is a suite of hardware and toolkits to support on-orbit sample processing)

SPICE – Smoke Point In Co-flow Experiment (the study examines smoke point phenomena in microgravity to improve understanding of soot emission from jet flames. Also, smoke point properties in co-flow environments are being studied with respect to nozzle diameter, flow velocity, and fuel consumption)

MELFI-EU (Minus Eighty-degree Laboratory Freezer for ISS Electronics Unit)

EVA hardware: 3 kg

EVA tools

Systems hardware: 135 kg

Environmental and Closed Loop Life Support Systems Hardware: CDRA Beds (Carbon Dioxide Removal Assembly – Orbital Replacement Units consisting of a CO_2 sorbent bed and desiccant bed)

Crew health care system components:
- Air quality monitor hardware
- Respiratory support pack hardware
- Fundoscope
- ARED (Advanced Resistive Exercise Device) ropes/lanyards
- T2 treadmill turbo cable

Flight crew equipment:
- 3.0 AH batteries
- Scope meter power adapter

Computer hardware: 8 kg
- Hard drives
- CD case
- Serial converter

Russian hardware: 0.3 kg
- TVIS (Treadmill Vibration Isolation System) gyro cable

External cargo: 273 (+ 100) kg – transported to the ISS in Dragon's trunk section

CRS-2 was the first Mission of Dragon that carried external cargo to the ISS using Dragon's Trunk Section. Dragon delivered two Heat Rejection Subsystem Grapple Fixtures (HRSGFs) to the ISS, which are two bars, each equipped with two Flight

Releasable Grapple Fixtures. These units will be used to provide grapple fixtures for Canadarm2 when moving ISS radiators in the event of a repair or replacement. The two HRSGFs were removed from Dragon's trunk by Canadarm2 during the docked mission. The two units will be temporarily stowed on the Payload Orbital Replacement Unit Accommodation on the Mobile Base System. During a future spacewalk, the HRSGFs will be attached to the S1 and P1 truss radiators in preparation for future maintenance operations.

Three weeks later, on March 26th, after bad weather at mission's end in the Pacific recovery zone kept it in orbit an extra day, Dragon returned to Earth with a full science load, splashing down in the Pacific, off the coast of Mexico's Baja Peninsula, five hours after leaving the ISS (Figures A11 and A12). Dragon brought back more than one tonne of science experiments and old ISS equipment (Panel A2). It was shipped to Los Angeles and then trucked to Texas for unloading. Earlier in the day, astronauts released the capsule from the end of the Space Station's robot arm. "Sad to see the Dragon go," astronaut Thomas Marshburn told Mission Control. "Performed her job beautifully. Heading back to her lair. Wish her all the best for the splashdown today."

Within hours of the splashdown, NASA had retrieved science samples and experiments that flew up with Dragon, including hundreds of flowering weeds. Mouse stem cells stayed behind on the Space Station, at the request of the Japanese researchers.

A11. Dragon being recovered after splashing down in the Pacific Ocean at the end of the CRS-2 mission. Courtesy: SpaceX

A12. Dragon recovered on deck. Courtesy: SpaceX

Panel A2. Dragon CRS-2 cargo manifest: Cargo down-mass

Total cargo down-mass: 1,210 kg
Total mass with packaging: 1,370 kg

Crew supplies: 95 kg
- Crew care, crew preference items
- Crew provisions
- Empty food containers

Experiment hardware: 660 kg

Canadian Space Agency:
Microflow
VASCULAR (Cardiovascular Health Consequences of Long-Duration Space Flight – studies the impact of long-duration spaceflight on the blood vessels of astronauts)

European Space Agency:
- ENERGY hardware
- Biolab LSM pump

JAXA experiments:
Medaka Osteoclast (Frozen Test Subjects) (the Medaka Osteoclast study

looks at the osteoclast activity and the gravity-sensing system of the vertebulate using Medaka fish)

Hair (hair samples taken from astronauts stored in MELFI)

Stem cells

HICARI (growth of homogeneous SiGe crystals in microgravity by the traveling liquidous zone method)

EPO (Education Payload Operations, a suite of small experiments that demonstrate basic principles of science, mathematics, technology, engineering, and geography)

MIB2 (Message in a Bottle 2, an outreach experiment consisting of a bottle that was exposed to the vacuum of space during a spacewalk and is "carrying" the vacuum of space back to Earth)

NASA experiments:

GLACIER

Double cold bags

HRP (Human Research Program equipment)

BCAT-C1 hardware (Binary Colloidal Alloy Test C1; "BCAT-C1 will probe three-phase separation kinetics and the competition between phase separation and crystallization in colloid polymer mixtures. This regime remains virtually uncharacterized in any type of material including molecular fluids or complex mixtures. BCAT-C1 takes advantage of a substantial opportunity to fill a gap in the knowledge of these fundamental processes. By examining the kinetics in seven samples of different composition, we intend to show that significant quantitative differences in kinetics occur even though the resulting phases are similar" (the official experiment description notes))

BRIC (Biological Research In Canisters)

Cell Bio Tech (Cell Biology Technology demonstration)

CGBA (Commercial Generic Bioprocessing Apparatus)

FCF supplies (FCF – Fluids & Combustion Facility)

Lego Bricks Payload (formerly NLO-Education-2, series of Lego kits that are assembled on orbit to demonstrate scientific concepts)

Microgravity science glovebox gloves

Sample collection kit hardware

SPHERES Equipment (SPHERES stands for Synchronized Position Hold, Engage, Reorient, Experimental Satellites and involves two satellites that are used inside the Space Station to provide a miniature test bed to study maneuvering capabilities and spacecraft measurement systems. Students developed the sequences that the two bowling-ball-sized microsatellites were put through. These two small satellites are used to study maneuver in space in miniature inside the ISS)

Vehicle atmosphere cabin monitor

EXPRESS rack stowage lockers

Surplus +4C ice bricks

EVA hardware: 38 kg
- Ion filter
- Spacesuit gloves
- Wire tie caddy
- REBA – Rechargeable EVA Battery Assembly
- ECOK – EMU Crew Options Kits
- CCAs – Communications Carrier Assemblies
- LCVGs – Liquid Cooling and Ventilation Garments

Systems hardware: 401 kg

Environmental and closed loop life support systems hardware:
- H2 sensor replacement unit – removed from CDRA
- Urine filter hose assembly
- Microbial check valve
- Control panel
- Pump separator
- Spent CDRA beds
- Ion exchange bed
- PBAs – Portable Breathing Apparatuses
- HEPA Filters (High-Efficiency Particulate Air Filters)
- Silver biocide kit

Crew health care system components:
- TEPC – Tissue Equivalent Proportional Counter
- Crank handle
- GSCs – Grab Sample Containers containing air samples
- CSA–CP – Compound Specific Analyzer–Combustion Products
- CSA–O2 – Compound Specific Analyzer–Oxygen
- RSP – Respiratory Support Pack
- RAMs – Radiation Area Monitors
- IV Supply Pack
- Medical equipment (IV Supply Pack, injection pack, oral medication pack)

Electrical power system:
- UOPs – Utility Outlet Panels
- RPCM III – Remote Power Control Module
- RPCM V – Remote Power Control Module

Other:
- Particulate filters
- Two double cargo transfer bags
- Crew command panel and permanent multipurpose module relocation equipment

Russian hardware: 16 kg
- Voltage and current stabilizer

LOOKING AHEAD: SPACEX FLIGHT MANIFEST

Customer	Vehicle arrival at launch site	Vehicle	Launch site
ORBCOMM – Multiple flights	2012–14	Multiple	Cape Canaveral
MDA Corp. (Canada)	2013	Falcon 9	Vandenberg
Falcon Heavy demo flight	2013	Falcon Heavy[1]	Vandenberg
SES (Europe)	2013	Falcon 9	Cape Canaveral
Thaicom (Thailand)	2013	Falcon 9	Cape Canaveral
NASA resupply to ISS – Flight 3	2013	F9/Dragon	Cape Canaveral
NASA Resupply to ISS – Flight 4	2013	F9/Dragon	Cape Canaveral
NSPO (Taiwan)	2013	Falcon 9	Vandenberg
AsiaSat	2014	Falcon 9	Cape Canaveral
AsiaSat	2014	Falcon 9	Cape Canaveral
NASA Resupply to ISS – Flight 5	2014	F9/Dragon	Cape Canaveral
NASA Resupply to ISS – Flight 6	2014	F9/Dragon	Cape Canaveral
NASA Resupply to ISS – Flight 7	2014	F9/Dragon	Cape Canaveral
Space Systems/Loral	2014	Falcon 9	Cape Canaveral
DSCOVR (US Air Force)	2014	Falcon 9	Cape Canaveral
CONAE (Argentina)	2014	Falcon 9	Vandenberg
DragonLab Mission 1	2014	F9/Dragon	Cape Canaveral
Asia Broadcast Satellite/Satmex	2014	Falcon 9	Cape Canaveral
Jason-3 for NASA	2014	Falcon 9	Vandenberg
Spacecom (Israel)	2015	Falcon 9	Cape Canaveral
NASA Resupply to ISS – Flight 8	2015	F9/Dragon	Cape Canaveral
NASA Resupply to ISS – Flight 9	2015	F9/Dragon	Cape Canaveral
NASA Resupply to ISS – Flight 10	2015	F9/Dragon	Cape Canaveral
Bigelow Aerospace	2015	Falcon 9	Cape Canaveral
DragonLab Mission 2	2015	F9/Dragon	Cape Canaveral
SES (Europe)	2015	Falcon 9	Cape Canaveral
CONAE (Argentina)	2015	Falcon 9	Vandenberg
Iridium – Flight 1	2015	Falcon 9	Vandenberg
Iridium – Flight 2	2015	Falcon 9	Vandenberg
Iridium – Flight 3	2015	Falcon 9	Vandenberg
NASA Resupply to ISS – Flight 11	2015	F9/Dragon	Cape Canaveral
NASA Resupply to ISS – Flight 12	2015	F9/Dragon	Cape Canaveral
STP-2 (US Air Force)	2015	Falcon Heavy	Cape Canaveral
Asia Broadcast Satellite/Satmex	2015	Falcon 9	Cape Canaveral
Intelsat	2015	Falcon Heavy	Cape Canaveral
Iridium – Flight 4	2016	Falcon 9	Vandenberg
Iridium – Flight 5	2016	Falcon 9	Cape Canaveral
Iridium – Flight 6	2016	Falcon 9	Vandenberg
Iridium – Flight 7	2017	Falcon 9	Vandenberg
Iridium – Flight 8	2017	Falcon 9	Vandenberg

[1] When the Falcon Heavy rocket is launched on its demo flight in late 2013, the vehicle will become one of the most powerful rockets ever launched (only the Saturn V, which sent

Americans to the Moon, has generated more power). The Falcon Heavy's 27 individual booster engines will generate 3.8 million pounds of thrust – enough power to lift the 3.1-million-pound rocket and its 117,000-pound payload into LEO. As you can see from the manifest, the Air Force has already hired SpaceX and its Falcon Heavy to send two satellites into orbit in 2015.

PAST MISSIONS

Customer	Launch	Vehicle	Launch site
NASA CRS-1	October 7, 2012	F9/Dragon	Cape Canaveral
NASA COTS – Demo 2/3	May 22, 2012	F9/Dragon	Cape Canaveral
NASA COTS – Demo 1	December 8, 2010	F9/Dragon	Cape Canaveral
Falcon 9 Inaugural test flight	June 4, 2010	Falcon 9	Cape Canaveral
ATSB (Malaysia)	July 13, 2009	Falcon 1	Kwajalein
Falcon 1 – Flight 4	September 28, 2008	Falcon 1	Kwajalein
US Government, ATSB and NASA	August 2, 2008	Falcon 1	Kwajalein
DemoFlight 2	March 20, 2007	Falcon 1	Kwajalein
DemoFlight 1	March 24, 2006	Falcon 1	Kwajalein

SPACEX MILESTONES

March 2002
SpaceX is incorporated

March 2006
First flight of SpaceX's Falcon 1 rocket

August 2006
NASA awards SpaceX $278 million to demonstrate delivery and return of cargo to the ISS

September 2008
Falcon 1, SpaceX's prototype rocket, is first privately developed liquid-fueled rocket to orbit Earth

December 2008
NASA awards SpaceX US$1.6 billion contract for 12 ISS cargo resupply flights

July 2009
Falcon 1 becomes first privately developed rocket to deliver a commercial satellite into orbit

June 2010
First flight of SpaceX's Falcon 9 rocket, which successfully achieves Earth orbit

December 2010

On Falcon 9's second flight and the Dragon spacecraft's first, SpaceX becomes the first commercial company to launch a spacecraft into orbit and recover it successfully

May 2012

SpaceX's Dragon becomes first commercial spacecraft to attach to the ISS, deliver cargo, and return to Earth

August 2012

SpaceX wins US$440 million NASA Space Act Agreement to continue developing Dragon to transport humans into space

October 2012

SpaceX completes the first of 12 official cargo resupply missions to the ISS, beginning a new era of commercial space transport

March 2013

SpaceX completes the second of 12 official cargo resupply missions to the ISS

Index

Made in the USA
Las Vegas, NV
19 December 2022

63271191R00136